Oil Shales

Miryam Glikson-Simpson

Oil Shales

A Complete Story

Springer

Miryam Glikson-Simpson
Private Consultant
Fossil Fuels, M. Glikson
Palmwoods, QLD, Australia

ISBN 978-3-030-60674-9 ISBN 978-3-030-60675-6 (eBook)
https://doi.org/10.1007/978-3-030-60675-6

Photograph by M. Glikson; Botryococcus alga in reflected light microscopy with UV excitation

This Springer imprint is published by the registered company Springer Nature Switzerland AG
The registered company address is: Gewerbestrasse 11, 6330 Cham, Switzerland

Acknowledgments

I like to thank Emeritus Professor Rod Simpson for support and encouragement, Senior Editor Mrs. Margaret Deignan, Production Editor Ritu Chandwani and the team for making this book a reality. I like to thank the reviewers, Prof. Bolhar, Witwatersrand University, and Dr. Uysal, University of Queensland, who have put the time and effort into reviewing the proposed manuscript.

I carried out all transmission electron microscopy (TEM-EDS), light microscopy and sample preparation at ANU (Australian National University during Ph.D. and Post Doc. Research years), in the ANU's Research School of Biological Sciences and the Department of Geology. C-isotope analyses I carried out in the CSIRO isotope laboratories in Sydney which were kindly made available to me.

Trace element analyses were carried out in the Department of Geology (ANU) by the Late Prof. Bruce Chappel, Ross Freemen and Elizabeth Webber. Major elements analysis was carried out by the ANU Analytical Unit in the Research School of Chemistry.

Introduction

What are oil shales, what makes them different from just 'shales' and how and why were some oil sales believed to be of commercial value, and are they really in the present. What are the health hazards and environmental implications of mining oil shale and processing shale oil. These will be discussed in a separate chapter. Most books and reports on oil shales are focusing on the commercial applications of the 'resource'. This book describes in great detail the components, organic and inorganic of oil shales, their sources and mode of residing within the deposits and their potential release from the oil shales, the cost of production and the cost to the environment and health: *the complete story.*

The term *Shales* as the name implies is referred to a group of laminated argillaceous sedimentary relatively impermeable rocks of varied texture, structure, mineralogy and geochemistry. The term *Oil Shales* implies a significant content of organic matter interspaced with the mineral matter. The information they provide on sources of the organic matter and mineral components and on depositional environments is invaluable. These rocks are therefore of great interest in understanding the genesis of organic-rich rocks and give a glimpse of the microorganisms that lived at the time.

Oil shales are commonly defined as finely laminated, fine-grained sediments, rich in organic matter which upon heating yields oil. This organic matter is *immature* on the coalification path, namely before reaching the stage of natural oil generation through burial and heat applied with increasing depth. The relatively high ash content (minerals) distinguishes oil shales from coal which is considered an organic rock. Furthermore, whereas a rock containing about two per cent organic matter could be a petroleum source rock, a rock with less than 20% organic matter could not be considered an oil shale, since the energy input to extract oil would exceed by far the energy output. In other words, oil shales were studied in terms of being a commercially exploitable resource. They have been utilized as a source of energy over the past one hundred years or so in France, Scotland, Sweden, Estonia, Spain, South Africa, Australia and China. All these 'oil shales' were organic-rich deposits, mostly consisting of lipid-rich algal remains; often a single planktonic alga being responsible for the deposit, such as *Botryococcus* in the torbanites of Australia, *Gloeocapsamorpha prisca* a mat forming planktonic alga in Kukersite the Estonian oil shale deposit, *Tasmanites* in the Tasmanian Tasmanite deposit and *Botryococcus* and other algae in the Green River oil shales. Not all of these commercially viable 'oil shales' in the past and present fit the term *SHALES*, as some of them such as torbanites are not laminated sediments in the structural and sedimentological definition of these rocks. The laminations in the Kukersite are formed from the fossilized algal mats interbedded with calcite. The torbanites are not laminated fine-grained sediments but oily (by touch), chunky black or dark brown organic deposits, and the Coorongite laminations are the result of layer upon layer of *Botryococcus* algal mats.

Several attempts have been made to classify oil shales on the basis of variation in the organic constituents, nature of mineral components, chemical composition, depositional environment and genesis. A few will be mentioned here.

A classification by Ozorov and Polozov (in Prien 1976) is productivity-driven based on oil yields whereby two main categories exist: *sapropelic* and *sapropelic humic*. The first category is composed of homogeneous high oil yielding oil shales with over 50% conversion of organic matter to oil and high heating value.

The second category comprises organic-rich oil shales with 15–50 % conversion to oil.

These authors further subdivided the two categories into four types on the basis of their commercial utilization:

1. Oil shales suitable for power production.
2. Oil shales suitable for gas and oil production.
3. Shales suitable for production of power and chemicals.
4. Shales suitable for oil production.

The drawbacks of this classification are evident. In order to be able to distinguish the type of oil shale for their commercial use, physico-chemical properties of the deposit need to be evaluated in the first instance to decide whether it is viable. The above classification does not answer these requirements.

Cane (1976) in discussing the nature of oil shales in the light of the breakdown of their chemical products proposed a relatively simple classification based on behaviour following oxidation: Two main categories were proposed, and these were further subdivided, each into two sub-categories:

Kerogen A: of algal origin, yielding no benzenoid acids.

A(i)—derived mainly from algal fatty acids

A(ii)—derived mostly from algal hydrocarbons

Kerogen B: yielding benzonoic acids; this kerogen contains humic contributors

A(i)—admixed with kerogen A(i)

A(ii)—admixed with kerogen A(ii)

Kerogen definition: Durand (1980): *The fraction of sedimentary organic matter which is insoluble in the usual organic solvents (chloroform, benzene, methanol-benzene) as opposed to the soluble fraction called bitumen*: in other words, '**Organic matter**' (OM) which is the preferred term in this book.

As a chemical classification, Cane's scheme is simple and useful, and indicates in a general way the source and nature of organic matter in a particular oil shale, which becomes potentially a general indicator of the expected properties of the oil and the spent shale after oil extraction. *Torbanite*, *Kukersite* and *Green River* oil shales yield no benzonoic acids on oxidation and therefore fall into category A, of algal origin. Torbanite falls into A(ii) and the two other oil shales into A(i).

Demaison and Moore (1980) divided oil shales on the basis of their depositional environment: in other words, an indirect method of identifying their petrographic characteristic in terms of expected minerals and trace elements. Four types of anoxic settings have been put forward, with examples from present-day analogues:

1. Large anoxic lakes
2. Anoxic silled basins
3. Anoxic layers caused by upwelling
4. Open ocean anoxic layers.

In terms of palaeo-climatic settings, type 1 forms in a uniform tropical environment; type 2 forms in temperate to warm conditions and includes intra-cratonic seas and anoxic basins on continental shelves. The fourth type is developed in conjunction with global warming and major transgressions, such as are known from Upper Jurassic to Lower Cretaceous. The Demaison and Moore (1980) classification has an advantage in having a broad aspect which can readily accommodate a wide range of oil shales. Furthermore, it provides basic information on the expected general petrological and chemical properties of the deposit to be analysed, in familiarity with the source material in these depositional environments. In defining the oil shales described in the present study in terms of the above classification, it will be seen that lacustrine deposits fit into type **1** and the torbanites into a sub-group of type **1**, namely small anoxic lakes. The Toolebuc oil shales and the Camooweal oil shales will be suggested as the product of type **2** depositional environment.

It is clear to me that a classification should be simple but informative at first instance about the basic properties of the deposit derived from the depositional setting which in turn will indicate whether the deposit is marine or fresh water and what type of microorganisms were likely to have made up the organic matter. These basic properties of the oil shale should give a preliminary prediction of the composition of the oil and residue upon heating. In other words, it should tell us something about the target output and the spent shale after retorting. It is however evident in view of the diverse composition of each oil shale deposit that none of the proposed classifications are of ultimate use to that purpose. To evaluate an oil shale deposit, it needs an array of analyses: X-ray diffraction to asses, the minerals making up the bulk of the deposit, elemental analysis to determine the organic carbon content, the hydrogen and oxygen and ash, which will indicate whether it is an oil shale. Reflected light microscopy of whole rock samples is important to obtain a fundamental picture of the type of organic and inorganic components and their relationships to one another, e.g. *Lamalginite* (Hutton 1987), indicating the OM being algae-forming lamellae, in a shale. Finally, trace element analyses are of great importance. In the present detailed study of selected oil shales, an attempt is made not only to provide an understanding of these deposits, but to point out the detrimental health and environmental effects of the residue left by these deposits during mining and after retorting. To achieve all that, a number of methodologies were applied which will be discussed in Chap. 1.

Understanding the Nature and Origin of Oil Shales: Problems and Methods

Understanding the nature of an oil shale involves determining the precursor organism or organic matter that makes up the particular deposit, and the degradation products that almost always accompany the precursor organism. Furthermore, the relationship between the organic and mineral fractions and the prevailing environmental conditions at the time of deposition are all important in assessing the properties and use of the deposits and evaluate their waste products. The main problem in understanding the nature of the organic constituents is the absence of primary structures visible to the naked eye. Even in viewing the 'amorphous' organic matter in transmitted light microscopy, or 'vitrinite-like' (vitrinite of very low reflectance/Ro) organic matter in reflected light microscopy, not many details are available. Other problems concern the nature of the organic matter and its trace element association. Where are the trace elements concentrated in relation to the organic matter types in different oil shales: Why specific trace elements are associated with particular oil shales, and with what fraction within the oil shale.

A common approach in the study of oil shales in the past has been to adopt one specific method and carry it out on numerous samples to elucidate the nature of the oil shale and its depositional environment. For example, light microscopy by itself, as a tool for understanding the source and nature of the organic matter in oil shales is limited by its optical level of resolution. In the present study, optical microscopy has been applied to organic matter concentrates as well as whole rock samples, using transmitted and reflected incident light as well as UV/fluorescence mode. Reflectance measurements (Glikson 1985, 2008, 2011) were carried out on the vitrinite-like (Ro $< 0.2\%$) organic matter and inertinite-like (higher Ro) OM in the Toolebuc and Camooweal, as well as the Mount Coolon oil shales to differentiate general organic groups lacking any structure visible in light microsopy.

The main difference between this study and other oil shale observations is the systematic use of transmission electron microscopy (TEM). TEM has shown structures within 'amorphous' organic matter that are beyond the resolving power of the optical microscope. Likewise, the close association of minerals with organic matter can be resolved only in TEM—EDS (Glikson et al. 1989; Glikson, 2001). Furthermore, the electron density gives a preliminary instant indication of the composition of the OM, as it is inversely proportional to reflected light microscope observation. Lipidic OM such as algae, immature pollen, spores and plant cuticles are relatively electron transparent compared to hardened, high Ro organic matter, humic acids and minerals which have higher degrees of electron density. Minerals which show in TEM as completely electron opaque particles are spot analysed by EDS-X-ray ('microprobe') which is attached to the transmission electron microscope.

One of the principal methods applied in studies of organic matter in sediments is carbon isotope analysis (Nissenbaum and Kaplan 1972; Peters et al. 1981; Rigby et al. 1981; Smith et al. 1982; Glikson and Taylor 1985; Glikson et al. 1989; Glikson et al. 2011). In the present study, C-isotope analyses have been applied to all the organic matter in the oil shales described in this book, as well as to the carbonates in the Toolebuc Formation oil shales ('Julia Ck.' oil shales). In the latter, oxygen isotope analysis was carried out on shell carbonates in an attempt to reconstruct past climate conditions in the depositional basin. Other methods applied in the present study include major element and trace element analysis. The former method served as an aid in tracing the sources of the organic matter and its degree of maturation. The latter method was employed to investigate the association of specific trace elements with particular oil shales. The general mineralogy of the oil shales was determined by X-ray diffraction.

References

Cane, R. F. (1976). The origin and formation of oil shale. In T. F. Yen & G. V. Chillingarian (Eds.), *Developments in Petroleum Science: Vol. 5. Oil shale* (pp. 27–60). Amsterdam: Elsevier.

Damison, G. J. & Moore, G. T. (1980). Anoxic environments and oil source bed genesis. *AAPG, 64*, 1179–1209.

Glikson M. & Taylor G. H. (1985). Cyanobacterial mats: Major contributors to organic matter in Toolebuc Formation oil shales. *Geological Society of Australia* (Special Publ.), *12*, 273–286.

Glikson, M. (1989). Botryococcus—A planktonic green alga, the source of petroleum through the ages: Transmission electron microscopical studies of oil shales and petroleum source rocks. *Organic Geochemistry, 14*(6), 595–608.

Glikson, M. (2001). The application of electron microscopy and microanalysis in conjunction with organic petrology to further the understanding of organic-mineral-association: examples from Mount Isa and McArthur Basins, Australia. *Coal Geology, 47*, 139–159.

Glikson, M., Duck, L. J., Golding, S. D., Hofmann, A., Bolhar, R., Webb, R., Baiano, J. F., Sly, L. I. (2008). Microbial remains in some earliest Earth rocks: Comparison with metanocaldococcus janaschii, a potential modern analogue. *Precambrian Research, 164*, 187–200.

Glikson, M., Hickman, A. H., Duck, L. J., Golding, S. D., Webb, R. E. (2011). Integration of observational and analytical methodologies to characterize organic matter in early archaean rocks: Distinguishing biological from abiotically synthesized carbonaceous matter structures. In S. D. Golding & M. Glikson (Eds.), *Earliest life on earth: Habitats, environments and methods of detection* (pp. 209–237). Springer.

Hutton, A. C. (1987). Petrographic approach to beneficiation of Australian oil shales. *Fuel, 66*: 314–318.

Nissebaum, A. & Kaplan, I. R. (1972). Chemical and isotopic evidence for the in situ origin of marine humic substances. *Limnology and Oceanography, 7*(4), 570–580.

Peters, K. E., Rohrback. B. G., Kaplan, I. R. (1981). Carbon and Hydrogen stable isotope variations in kerogen during laboratory-simulated thermal maturation. *AAPG Bulletin, 65*(3), 501–508.

Prien, C. H. (1976). Survey of oil shale research in the last three decades. In: T. F. Yen & G. V. Chillingarian (Eds). *Developments in Petroleum Science: Vol. 5. Oil shale* (pp. 235–267).

Rigby, D., Batts, B. D., Smith, J. W. (1981). The effect of maturation on the isotopic composition of fossil fuels. *Organic Geochemistry, 3*, 29–36.

Smith, J. W., Gould, K. W., Rigby, D. (1982).The stable isotope geochemistry of Australian coals. *Organic Geochemistry, 3*, 111–131.

Contents

Chapter 1
Methods of Study

Abstract The methods of study used to elucidate the nature and composition of the oil shales described in this book are outlined and described briefly. The use of Transmission Electron Microscopy (TEM) was applied to the organic matter concentrated from the rock samples. TEM enabled the observation of microbes and ultra-fine structures not visible when using conventional microscopical techniques. Furthermore, using Electron Dispersive Spectrometry (EDS) coupled with TEM enabled the identification of minerals residing within, and associated with specific organic entities. The defining of organic elements such as microbial filaments, algal remains and degraded OM were confirmed by using staining techniques e.g. lipidic/algal/spore-pollen exines do not absorb metal stains such as uranyl-acetate or lead citrate used in the experiments. Humic acids, an important part of all organic matter in sediments were identifiable in TEM.

Oil shales were chosen from different geological formations, belonging to different geological ages, varied depositional environments and organic matter precursors. Mineral composition and trace element concentrations were highlighted in the various oil shales. Some will prove to be oil shales by name but not a commercial resource. The following problems were researched; the nature of organic-mineral associations, the source of certain trace elements and metals, the significance of the humic acid fraction of the organic matter, maturation stage of the organic matter and potential oil yields. Major elements and Trace element analyses were carried out on organic matter, whole rock samples as well as spent shales. The latter are discussed in chapter VII. Humic acids were separated from the rest of OM and viewed in TEM to enable their recognition in all oil shales.

1.1 Major Elements

Carbon and hydrogen make up most of the organic matter in all types of organic deposits, whereas oxygen is the next most abundant element, particularly in the humic acid fraction of the organic matter. Nitrogen and Sulphur vary in abundance

M. Glikson-Simpson, *Oil Shales*, https://doi.org/10.1007/978-3-030-60675-6_1

according to the type of organic matter, and inorganic sulphur occurs as pyrite. High nitrogen content is usually obtained from humic acid-rich samples associated with degradation products of the precursor organisms. Carbon is progressively concentrated with an increase in the degree of maturation. Different organic entities differ in their maturation paths. Sulphur content will affect production and processing of the oil shales.

The classic Van Krevelen (Taylor et al. 1998) diagram developed for coals is adopted here for oil shales. By expressing the ratios between the elements carbon, hydrogen and oxygen, three types of OM (kerogen) have been designated (Tissot and Welte 1978) on the basis of concentration of the elements relative to each other. Type I OM (kerogen) has a high hydrogen content (H/C1.5 or more) as a result of highly aliphatic hydrocarbons which reflect the algal origin of this type of organic matter. This OM is low in oxygen (O/C less than 0.1) and is highly desirable as a source rock for oil. As humic matter increases the H/C decreases, O/C increases and hydrocarbon potential of the source rock decreases. However, with advancement of organic matter maturation and the increase in temperature that OM is subjected to during its burial, the differences between the OM types become masked. Thus, the carbon, hydrogen and oxygen content of an oil shale may indicate its potential oil yield.

1.2 Trace Elements

Trace element analysis was carried out on OM concentrates as well as powdered whole rock samples using X-ray fluorescence spectroscopy (Norrish and Chappel 1979). The association of trace metals with organic matter raises questions regarding the mode of metal concentration, the type of organic matter or particular organisms involved with the specific trace elements. Clay minerals play an important role in the sedimentary environment through their absorbing capacities and affinities towards metals. Questions arise as to why were some metals absorbed by, organic matter and others on clay minerals, and why are some metals abundant in a particular fraction of organic matter such as humic acids.

The affinity and relationship of particular groups of organisms to trace metals has been the subject of numerous studies: Brown and Lester (1979) demonstrated the role of bacteria and flocculation and subsequent precipitation of metals on organic matter. According to those studies, the metal may be removed by absorption on to sites on bacterial extra-cellular polymers, on bacterial cell walls, or accumulated in the cytoplasm of bacterial cells.

Studies by Beveridge and Murray (1980) using atomic absorption and TEM on the uptake and retention of metals in *Bacillus subtilis* demonstrated that there were preferred sites for the absorption and retention of various cations. These authors found that Mg, Fe, Cu, Na, and K were deposited in the wall of the bacterium in significant quantities; Mn, Zn, and Ca retained in intermediate amounts, and only small amounts of Hg, Sr, Pb, and Ag. The elements Li, Ba, Co and Al were not taken into the bacteria's wall. Elements of higher atomic number could be detected as electron dense particles, diffusely staining walls. These studies have further elucidated the various factors influencing the uptake of certain metals by observing that the selection of one ligand over another is controlled mainly by the size of the cat-ion, electronegativity and the arrangement or size of hydration shells. Furthermore, in the studies of the sites of metal deposition in bacterial cell walls, it was concluded that carboxyl groups are of major importance in providing sites for metal deposition. The role of bacteria in concentrating metals and thus contributing to the formation of ore deposits in the geological past has been documented; e.g. Degens and Ittekkot (1982) who identified coccoid bacteria in electron dense particles from the Cretaceous Messel oil shale.

1.3 Carbon Isotopes

Stable isotope analysis, which is highly useful for the identification of organic matter source and elucidating the diagenetic history of both organic and inorganic components, was not in the past widely applied to oil shales. In dealing with organic matter, carbon-isotope variations can be utilized in the following way (Stahl 1977):

1. The tracing of organic source material
2. Recognition of bacterial degradation products
3. Recognising autochthonous and allochthonous fractions of organic matter.

Almost all organic matter incorporated in oil shales from the upper Palaeozoic has been directly or indirectly produced by photosynthesis. The assimilation of carbon by organisms arises from two sources: Atmospheric CO_2 (molecular CO_2) and/or HCO_3 in water. Thus carbon-isotope values will be related to the carbon source utilised by the organism. It is known that atmospheric CO_2 is depleted in ^{13}C relative to the inorganic carbon pool of the oceans. Fractionation therefore occurs in the hydration stage from dissolved CO_2 to HCO_3 and not from gaseous CO_2 to dissolved CO_2, Which means that in isotopic equilibrium, the CO_2 of the atmosphere should have the same ^{13}C as the dissolved CO_2 in the underlying water body, whereas the HCO_3^- will be enriched in ^{13}C by some 5 ppt. Since in normal sea water, with comparatively high pH most of the dissolved inorganic carbon is in the form of HCO_3^-, it will be isotopically heavier than carbon in fresh water bodies, where it is mostly in CO_2 form.

It has been stressed (Degens 1969) that maximum fractionation is obtained at low pH and low water temperature, with a moderate growth rate of the organism at high CO_2 concentration. The importance of the temperature factor was also emphasised by Deuser et al. (1968) in comparing ^{13}C values of plankton from different parts of the oceans over a range of temperature from 0° to 30°. On the other hand, Rau et al. (1982) argues against temperature control of ^{13}C in plankton by comparing isotopic compositions of arctic and ant-arctic plankton from the same latitude and temperature. Their study obtained different trends in the two hemispheres. They concluded that a 'strict temperature—plankton/^{13}C relationship common to all oceans does not exist and that other factors determining plankton ^{13}C are involved, such as dark CO_2 fixation reported from studies in Antarctic waters' (Horrigan 1981). The importance of temperature effect in isotope fractionation is noted (Galimov 1980) but can be masked by other factors. The same study stressed thermodynamics as responsible for the carbon isotope fractionation. Galimov developed formulae to calculate the thermodynamic isotope exchange properties of organic compounds by finding 'the partition function ratios of isotopic forms' which he refers to as the factor-characterising the carbon atoms as a function of the type of bond in the molecule. Thus, his study obtained correspondence between experimentally measured ^{13}C and theoretically calculated values; his conclusion being that "*if biological isotope effects are thermodynamically ordered, the carbon isotopes will be regulated not only between biomolecules but also within them. Bearing in mind the different molecular arrangement of various organic compounds within an organism, different isotopic values would be expected within different compounds*". Interestingly, the experimental work of Londry and De Marais (2003) has proved that. This will be discussed in more detail in the following chapters.

Processes involved in photosynthesis and metabolism occurring at various stages of the algae/plant growth are associated with isotope fractionation: The carboxylation step where CO_2 is enzymatically fixed leading to the formation of phosphor-glyceric acid was considered by Park and Epstein (1960, 1961) to be the major metabolic step involving the greatest fractionation (up to 17 ppt). The biosynthesis of amino acids, being complex, provides a step for carbon isotope fractionation, as was demonstrated by Abelson and Hoering (1961). The latter authors demonstrated a range of 17 ppt for ^{13}C between the various amino acids, and within the amino acid molecule itself, where the carboxyl group is enriched in ^{13}C by as much as 20 ppt relative to the remainder of the molecule. It was shown by Park and Epstein (1960, 1961) that lipids tend to be depleted in ^{13}C relative to the rest of the plant; a significant observation when interpreting isotopic values of ancient sediments, considering that lipid-rich parts of plants and in particular algae are selectively preserved.

Differences in isotope values of source organic matter are preserved in fossil organic matter on a reduced scale (Galimov 1980). Organic matter in recent sediments has isotopic compositions close to those of living organisms. Upon the death of an organism and onset of biodegradation, preferential elimination of ^{13}C-rich compounds such as carbohydrates and proteins will bring about enrichment in ^{12}C. Later, decarboxylation further lowers the ^{13}C content. Biodegradation followed by diagenesis will therefore result in 'lighter' residual organic matter with time. Exception to this 'rule' seem to be in some Permian coals (Wetering et al. 2013; Aggarval et al. 2020) where very 'heavy' isotopic compositions are present. Algal dominance at those times was suggested by Glikson-Simpson (2020) as a reason for isotopically 'heavy' values in the Permian coals. Isotopically 'heavy' OM was obtained from *Botryococcus* dominated Coorongite (Glikson 1984) which is associated with calcareous material. Turner and Fritz (2011) showed enrichment in ^{13}C of interstitial waters which can be rich in calcite up to saturation. Algae would have used C from Bi-carbonate instead of CO_2 for metabolism.

The isotopic value of coals falls close to the mean value of wood, $^{13}C = -25$ ppt, and there is no correlation between isotopic composition and geologic age of coals. Similar conclusions were reached by Rigby et al. (1981) from their extensive studies of maturation effect on the carbon isotopic composition of fossil fuels. McKirdy and Powell (1974) maintained that the carbon isotopic composition of organic matter in Proterozoic to middle Cambrian sediments had changed due to thermal effect. Heating, stressed by Hower (1993) is a crucial factor in the isotopic difference between species.

Thermal effect in Archaean sediments was demonstrated in samples from the Pilbara Craton of Western Australia and the Barberton Greenstone Belt of South Africa where high temperatures brought on enrichment in ^{13}C in the 3.5 billion-year old recovered organic matter. The OM gave a range of Ro values with the minimum of 1.2–1.8% which is equivalent to 170–200 °C (Glikson et al. 2008, 2011). These ancient sediments were deposited around deep-sea hydrothermal vents.

Of relevance to the present study are the bio-degradational changes imposed on the organic matter at the times of deposition. Bacterial decomposition leads to breakdown of proteins and carbohydrates and subsequent concentration of lipidic compounds. Since lipids are depleted in ^{13}C relative to the rest of the organic compounds, their increase with depth will have an immediate impact on the carbon isotope composition.

Isotopic studies of the humic fraction of the organic matter in sediments carried out by Nissenbaum and Kaplan (1972) are of importance in establishing their source in relation to the rest of the organic matter. In the present study carbon isotope data will be discussed in conjunction with light microscopy as well as TEM of the organic matter. Thus, the combined methodologies enable to distinguish between in-situ-formed autochthonous, and allochthonous organic matter derived from a source outside the depositional environment of the parent material. Therefore, humic acids were extracted from the bulk of OM and analysed separately.

Methods used in determining C-isotopic composition were as reported in earlier studies.

The carbon isotope data are reported as ‰ difference in absolute isotopic ratios between the investigated sample and the standard PDB, expressed as follows:

$$\delta^{13}\text{C sample} = \frac{^{13}\text{C}/^{12}\text{C sample} -^{13}\text{C}/^{12}\text{C standard}}{^{13}\text{C}/^{12}\text{C standard}} \times 10^3$$

Calcareous material of selected samples has been analysed for their isotopic composition with the aim of tracing their carbon source. $CaCO_3$ from benthic shells was compared to calcite closely associated with organic matter.

C-isotope values are used in this study as yet another parameter aiding in classifying oil shales, comparing them and relating them to one another. Correlations between parameters such as $\delta^{13}C$ and TEM observations throw light on the composition of organic matter in the oil shales, and therefore help in understanding these deposits and their diagenetic history.

1.4 Light Microscopy

Microscopical observations—were carried out in both; reflected and transmitted light; reflectance measurements (Ro), point counting of macerals (vitrinite, vitrinite-like, inertinite), in accordance with established techniques (Taylor et al. 1998) on samples of all the organic matter in the oil shales reported in this book. Ro conversion to temperatures follows Aizawa's (1990) calibration chart.

1.5 Transmission Electron Microscopy (TEM)

Transmission Electron Microscopy (TEM) is the unique and generally uncommon methodology used in this book; it has been applied to the organic matter in all the oil shales described in the following chapters. The method was initially applied by the author to sedimentary organic matter and is outlined in a number of publications (Glikson and Taylor 1985; Glikson 2001; Duck et al. 2007; Glikson et al. 2008). TEM is used in this book to show structures not visible at lower magnification of conventional microscopy and therefore of great value in identifying microbial remains and ultra-structures of algae, and often provides the only evidence and credence to the identification of sources of the OM.

Thus, TEM is a method that presents the ultimate means to the visual identification of precursor organisms in oil shales where no structures are discernible to the eye in light microscopy. The use of EDS coupled with TEM enables direct identification of metal particles associated with the OM.

All the methodologies mentioned above are used in this study as parameters in evaluating the origins and sources of biological entities and as tools in understanding the diagenetic history of oil shales.

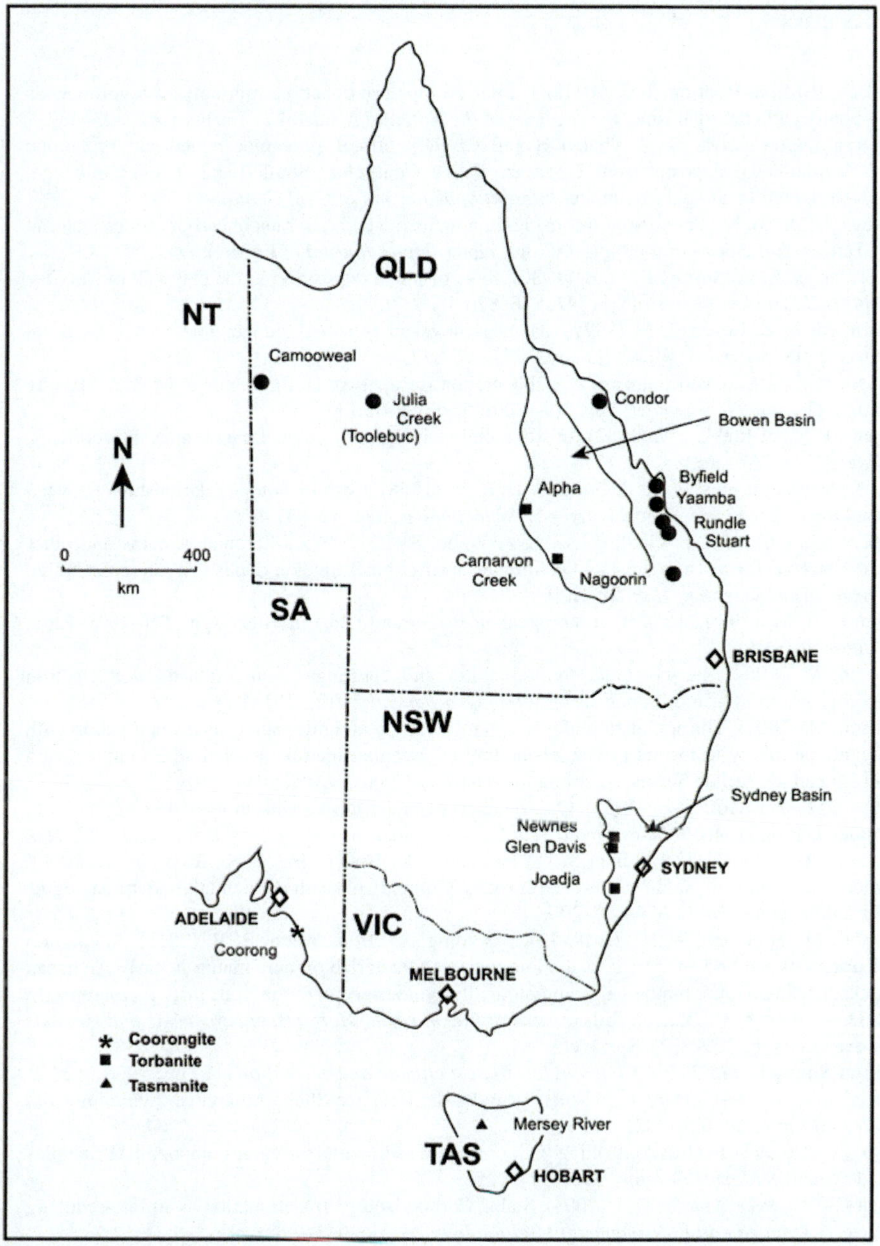

Fig. 1.1 Key map Australian oil shale deposits. *Source* Glikson (1984, unpublished)

References

Abelson, P. M., & Hoering, T. C. (1961). Carbon isotope fractionation information of amino acids by photosynthetic organisms. *Proceedings of the National Academy of Sciences, 47,* 623–632.

Aggarval, N., Aggarval, S., & Thakur B. (2020). Palynofloral, playnofacies and carbon isotope of Permian coal deposits from Godavari Valley Coalfields, South India: Insight into age, palaeovegetation and palaeoclimate. *International Journal of Coal Geology.*

Aizawa, J. (1990). Palaeotemperatures using fluid inclusion and coal rank of carbonaceous material of Tertiary formations in south-est Kyushu, Japan. *Japan Journal of Earth Science, 85,* 145–154.

Beveridge, T. J., & Murray, R. G. E. (1980). Sites of metal deposition in the cell wall of *Bacillus subtilis*. *Journal of Bacteriology, 141,* 876–870.

Brown, M. J., & Lester, J. N. (1979). Metal removal in activated sludge: The role of bacterial extracellular polymers. *Water Research, 13,* 817–837.

Degens, E. T. (1969). Biochemistry of stable carbon isotopes. In G. Eglongton & M. T. J. Murphy (Eds.), *Organic Geochemistry* (pp. 304–309). Springer Berlin.

Degens, E. T., & Ittekkot, V. (1982). In situ metal-staining of biological membranes in sediments. *Nature, 298,* 262–268.

Deuser, W. G., Degens, E. T., & Guillard, R. R. L. (1968). Carbon isotope relationships between plankton and sea water. *Geochimica et Cosmochimica Acta, 32,* 657–660.

Duck, L. J., Glikson, M., Golding, S. D., & Webb, R. E. (2007). Microbial remains and other carbonaceous forms from the 3.24 Ga Sulphur Springs black smoker deposit, Western Australia. *Precambrian Research, 154,* 205–220.

Galimov, E. M. (1980). $^{13}C/^{12}C$ in kerogen. In B. Durand (Ed.), *Kerogen* (pp. 271–299). Paris. Edition Technip.

Glikson, M. (1984). Furtherstudies on torbanites and coorongite using transmission electron microscopy and C-isotope analysis. *Organic Geochemistry, 7*(2), 151–160.

Glikson, M. (2001). The application of electron microscopy and microanalysis in conjunction with organic petrology to further the understanding of organic-mineral association: Examples from Mt Isa and McArthur Basins, Australia. *International Journal of Coal Geology, 47,* 139–159.

Glikson, M., & Taylor, G. H. (1985). Cyanobacterial mats: Major contributors to organic matter in Toolebuc Formation oil shales. *Geological Society of Australia Special Publication, 12,* 273–286.

Glikson, M., Duck, L. J., Golding, S. D., Hofmann, A., Bolhar, B., Webb, R. E., et al. (2008). Microbial remains in some earliest Earth rocks: Comparison with a potential modern analogue. *Precambrian Research, 164,* 187–200.

Glikson, M., Hickman, A. H., Duck, J. L., Golding, S. D., & Webb, R. E. (2011). Integration of observational and analytical methodologies to characterize organic matter in early Archaean rocks; Distinguishing biological from abiotically synthesized carbonaceous matter structures. In S. D. Golding & M. Glikson (Eds.), *Earliest life on earth; Habitats, environments and methods of detection* (pp. 209–237). Springer.

Glikson-Simpson, M. (2020). *Coal-window to past climate and vegetation* (170 pp.). Springer.

Horrigan, S. G. (1981). Primary production under the Ross Ice Shelf, Antarctica. *Limnology and Oceanography, 26,* 378–382.

Hower, J. M. (1993). Factors controlling ^{13}C contents of sedimentary organic compounds: Principles and evidence. *Marine Geology, 113*(111), 125.

Londry, K. L., & De Marais, D. J. (2003). Stable Carbon Isotope fraactionation by sulfate-reducing bacteria. *Applied and Environmental Microbiology, 69*(5), 2942–2949.

McKirdy, D. M., & Powell, T. G. (1974). Metamorphic alteration of carbon isotopic composition in ancient sedimentary organic matter: New evidence from Australia and South Africa. *Geology,* 591–595.

Nissenbaum, A., & Kaplan, I. R. (1972). Chemical and isotopic evidence for the in situ origin of marine humic substances. *Limnology and Oceanography, 17*(4), 570–580.

Norrish, K., & Chappel, B. W. (1979). X-ray fluorescence spectrometry. In J. Zussman (Ed.), *Physical methods in determinative mineralogy*. London: Academic Press.

Park, R., & Epstein, S. (1960). Carbon isotopes fractionation during photosynthesis. *Geochimica et Cosmochimica Acta, 21,* 110–126.

Park, R., & Epstein, S. (1961). Metabolic fractionation of ^{13}C and ^{12}C in plants. *Plant Physiology, 36,* 133–138.

Rau, G. H., Sweeney, R. E., & Kaplan, I. R. (1982). Plankton 13C:12C ratiochanges with latitude: differneces between northern and southern oaceans. *Deep Sea Research, 29,* 1035–1039.

Rigby, D., Batts, B. E., & Smith, J. W. (1981). The effect of maturation on the isotopic composition of fossil fuels. *Organic Geochemistry, 3,* 29–36.

Stahl, W. J. (1977). Carbon and nitrogen isotopes in hydrocarbon research and exploration. *Chemical Geology, 20,* 121–149.

Taylor, G. H., Teichmuller, M., Davis, A., Diessel, C. F. K., Littke R., & Robert, P. (1998). *Organic Petrology* (704 pp.). Gebruder Borntraeger, Berlin, Stuttgart.

Tissot, B. P., & Welte, D. H. (1978). *Petroleum formation and occurrence. A new approach to oil and gas exploration* (538 pp.). Berlin: Springer.

Turner, J. V., & Fritz, P. (2011). Enriched ^{13}C composition of interstitial water in sediment of fresh water lake. *Organic Geochemistry, 1993,* 265–276.

van Krevelen, D. W. (1961). *Coal Typology, Chemistry, Physics, Constitution.* (p. 514). Amsterdam: Elsevier.

Wetering, N., Esterle, J., & Baublys, K. (2013). Decoupling $\delta^{13}C$ response to palaeoflora cycles and climatic variation in coal: A case study from the late Permian Bowen Basin, Queensland, Australia. *Palaeogeography, Palaeoclimatology, Palaeobotany, 386,* 165–179.

Chapter 2
Unique Algal—Dominated Oil Shales

Abstract The organic matter in the oil shale deposits described in this chapter are dominated by algal remains. The organic matter in the torbanites is solely composed of the alga *Botryococcus braunii* Kutzig. The unusual properties of this alga to secrete hydrocarbon oils has made it a prime contributor to the oil-rich oil shales such as torbanites, Green River and in localised various other organic-rich deposits. The alga is associated with flooding of Permian swamps and other fresh water ponds and lakes throughout the geological history. Many studies have been published over the years concerning this alga. In the past 10 years experimental research and genetics targeting this alga have been numerous; the aim being a potential biofuel. Torbanite which is mostly made up of this alga, when heated produces 600–900 l/ton of oil. The TEM of the torbanites highlights the lipid bodies of the alga which are the source of oil. The Coorongite deposits forming today in lagoons are shown in TEM to be composed of layer upon layer of the alga *Botryococcus*. Experimental studies (Iyer et al. in Asian J. Microbiol. Biotechnol. Environ. Sci. 14:523–535, 2012) demonstrated the Coorongite *Botryococcus* being a stage of the fresh water alga when salinities are increased. Tasmanite, and Kukersite also algal dominated high oil yielding oil shales are the result of accumulated algal cysts. The *Tasmanites* cyst belongs to the algae group Prasnophycaea. The deposit of Late Carboniferous age was mined in Tasmania between 1910 and 1925. Tasmanite yielded up to 300 l/ton with an average of 130 l/ton. Kukersite the main energy source of Estonia has been mined in the past and is being mined presently. The source of the oil is a colonial alga by the name of *Gloeocapsomorpha prisca* which closely resembles the living alga *Entophysalys*.

2.1 Torbanites and Coorongite

Australian organic-rich torbanites have been recognised as having the highest oil yield of all oil shales as indicated by their highest H/C. Torbanites were mined in Australia until the early 1950s.

Glen Davis in New South Wales was the first oil shale mining town, established in 1881. Mining operations began with major mining enterprise in 1938. The town

M. Glikson-Simpson, *Oil Shales*, https://doi.org/10.1007/978-3-030-60675-6_2

Table 2.1 Maceral composition

% Maceral composition of torbanites (by point count)

	Alginte	Vitrinite	Vitrinite-like	Inertinite
TC-1	73	9	13	5
TA-1	50	23	23	4
Glen Davis	50	10	38	2

developed during the 1940s as a main oil producer. The operation closed down in 1952 when petroleum imported was cheaper than retorting from oil shales. All other retorts in New South Wales torbanite mines stopped operation at that time.

Torbanites are organic-rich lacustrine deposits associated with Permian coal measures in the Sydney and Bowen Basins. The deposits are dominated by the remains of the green alga *Botryococcus braunii* (Figs. 2.1, 2.2 and 2.3). The geochemistry of *Botryococcus* has been the subject of numerous studies; (Derenne et al. 1990; Derenne 1997). Allan et al. (1979) who were first to note 'stringers of a low-reflectance, uniform grey material of unknown origin in Torbanites'. These are visible in TEM of the present study (Figs. 2.7 and 2.8). Lipid bodies, source of hydrocarbons visible in TEM of *Botryococcus* were studied by Largeau et al. (1980) and Metzger and Largeau (2005). The present study showed that alginite (*Botryococcus* remains) made up 90% of the organic matter. The rest contained low reflectance vitrinite-like OM and inertinite-like OM. The H/C atomic ratio of Torbanite when plotted on a Van-Krevelen diagram placed this 'oil shale' in type I 'kerogen'.

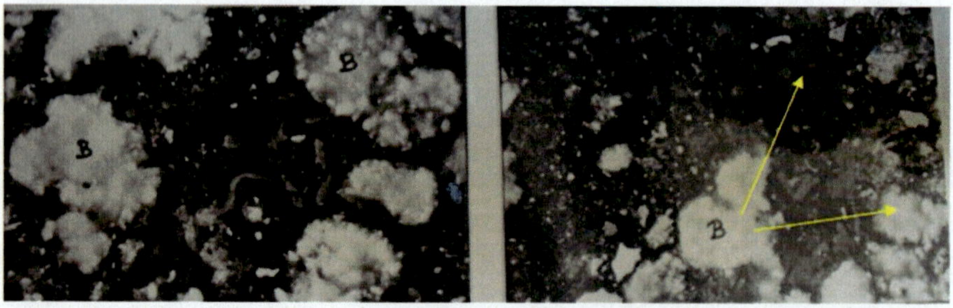

Fig. 2.1 *Botryococcus* colonies in reflected light microscopy. Photo taken at ×100 mag

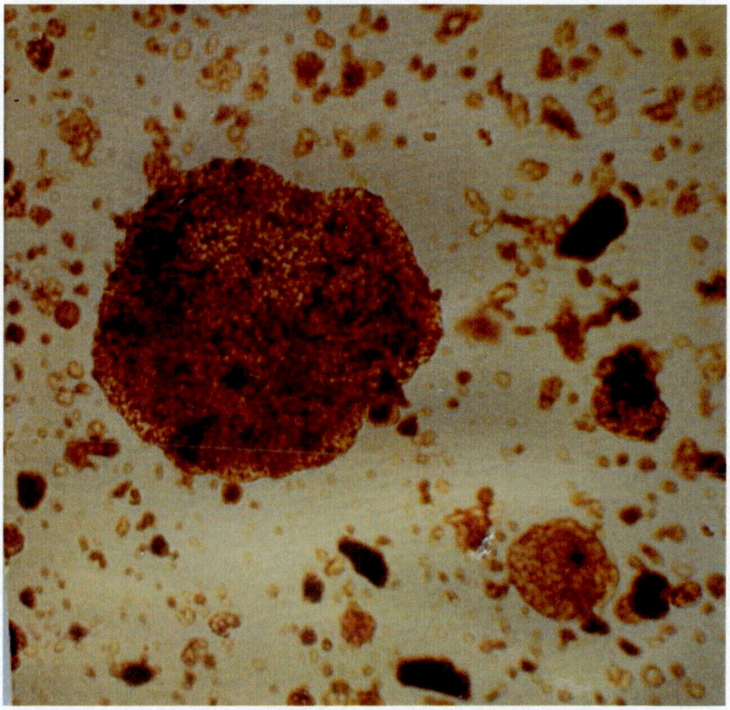

Fig. 2.2 *Botryococcus* colonies and dispersed single cells in transmitted light microscopy (above) in OM concentrate. ×500

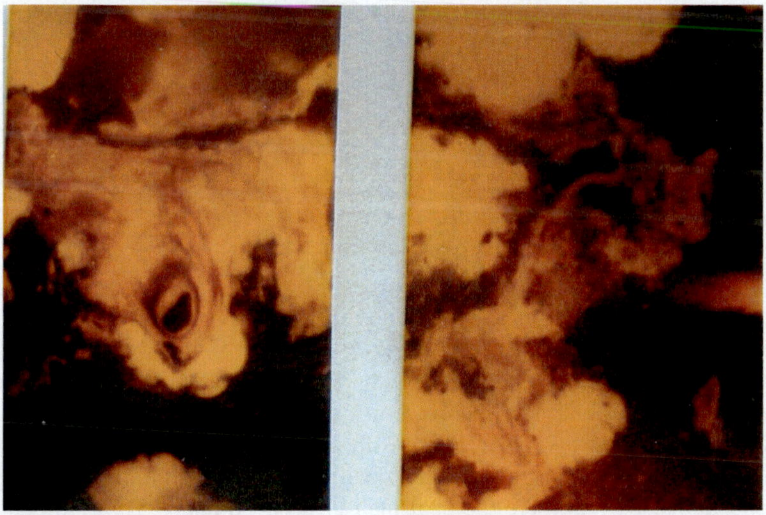

Fig. 2.3 *Botryococcus* algal cells in reflected light with UV excitation. Yellow fluorescing parts before oil generation; brown parts after oil generation. ×500

MAP SHOWING THE RELATION BETWEEN THE DISTRIBUTION OF KNOWN TORBANITE DEPOSITS AND THE PROBABLE AREA COVERED BY FRESHWATER SWAMPS OF THE UPPER COAL MEASURES, NEW SOUTH WALES

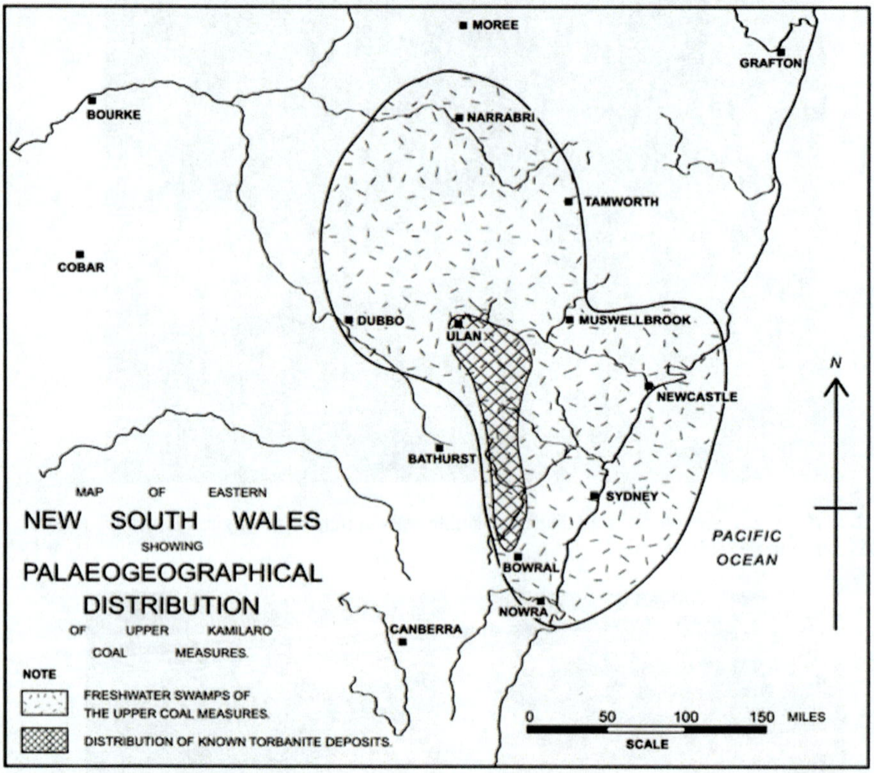

From: Dulhunty, 1944

The origin and formation of torbanites and their analogy with Coorongite had been reported by Dulhunty and his co-workers (1941, 1943, 1944). Dulhunty focused on the torbanites of Sydney Basin, New South Wales. Connah (1964) described the Torbanites from Queensland. The hydrocarbon-rich nature of these deposits has been noted. The Torbanite deposits in New South Wales occur in Upper and Lower Permian coal measures. The torbanites in the upper coal measures are confined to the edge of the depositional basin, where the coal measures are thinnest. These torbanites are confined to the Newcastle stage of the Permian, during which a depositional environment favourable for their formation prevailed; such as inflow of fresh water into depressed areas in the western part of the basin that supported ponds ideal for algal bloom. The flooding of these Glossopteris swamps in the Permian signified a climatic change; the formation of lake systems and algal blooms (Glikson-Simpson 2020).

The Alpha torbanite of the Bowen Basin is also associated with Permian coal seams and is identical in appearance to the torbanites of the Sydney Basin. The Queensland torbanite forms a 3 km long and 1 km wide lens about 1 m thick within a thick coal seam at the edge of the basin. Another torbanite deposit occurs along the western margin of the Bowen Basin and is exposed along the Carnarvon Creek. These large fresh water ponds within the thick Permian coal deposits are evidence to fluctuations in the climate and water regimes during the Permian.

The Alpha torbanite deposit of Permian age was described by Madre (1987) in detail. The same study noted the relatively small lens-like occurrence of torbanite within a more extensive pod of cannel coal. The deposit is located on the south-eastern edge of the Galilee Basin. The torbanite-cannel coal of the Alpha deposit has been assessed commercially (Madre 1987) having a very high oil yield, namely up to 622 l/ton of the pure torbanite part, and 259 l/ton of the cannel coal part.

In experiments of simulation of hydrocarbon generation from torbanite and brown coal over six years (Saxby et al. 1986), the torbanite showed maximum oil generation at 300 °C. Quantity of oil generated from torbanite exceeded 600 l/ton.

When conducting microscopic studies of polished rock surfaces of Torbanite in ultra-violet light (UV) with blue filter excitation, the brightly fluorescing *Botryococcus* colonies are the prominent visible features (Fig. 2.3). However, there are other parts of the sample that exhibit brown fluorescence under blue light excitation. These are algal cells and exines of spores/pollen which generated oil earlier (Glikson et al. 1992). The brown fluorescing organic matter, in close association with the bright yellow algal colonies, displays concentric structure following algal walls. In addition to the brown and yellow colours generated under UV light (blue filter), there are some areas devoid of any fluorescence. Such areas can make up to 10% of the volume of the torbanite sample. In incident reflected light (Fig. 2.2), very low reflectance vitrinite-like OM was seen to make up the very dark areas devoid of fluorescence between parts of the brown fluorescing 'envelopes' around bright yellow parts of the *Botryococcus* colonies (Fig. 2.3). In TEM (Fig. 2.6) these structures show as darker, electron denser material surrounding electron transparent material. It is also noteworthy that the darker concentric structures in TEM are seen to be associated with humic acids, or often only partly broken up OM grading into humic acids. Suggesting it being partially 'degraded algal residue'. Where pure algal remains are encountered in TEM they appear as a homogenous material of low electron density (Fig. 2.8).

TEM of torbanites revealed, in addition to *Botryococcus*, bundles of filamentous microbial entities, lipid bodies and degraded OM (Figs. 2.6 and 2.7). The filamentous structures closely resemble *Methanobacterium formicium*. Spherical entities with fuzzy outline (Fig. 2.4) resembling coccoid methanogens were also observed.

Living methanogens fluoresce only when oxidised (Zeikus 1977), hence blue fluorescence of these bacteria when exposed to air. However, they do not fluoresce when preserved under reducing conditions. This may explain lack of fluorescing properties of some of the areas between the fluorescing *Botryococcus* colonies. The

Fig. 2.4 TEM: probable methanogenic bacteria. ×140,000

highly anoxic environment into which the algal accumulations sank would have supported only anaerobes such as methanogens (Fig. 2.4) and sulphate reducing bacteria.

Methanogenic bacteria, being chemoautotrophs, are able to use hydrogen as a source of energy and CO_2 for cellular carbon synthesis. Therefore, a symbiotic relationship between methanogens and other decomposing organisms likely existed. Such relationships have been reported in an earlier study (Reddy et al. 1972). Thus, fermentation of carbohydrates leads to release of hydrogen, CO_2, ethanol, formic and acetic acids (Hobson et al. 1974), providing a suitable medium for the methanogens. A fraction of the carbohydrates may have reacted with amino acids (degradation products of the alga) in a melanoidin-type reaction (Hedges 1978) forming the humic acids observed in TEM of the samples. This is supported by the close isotope compositions of the humic acid fraction and rest of OM (Table 2.2). The ultra-structure of humic acids has been established by resin-embedding and ultra-sectioning the alkali-extractable fraction (Fig. 2.5).

Coorongite is basically made of Botryococcal mats with a considerable input of diatoms (Plate Fig. 2.10). *Botryococcus*—dominated algal mats have been shown in previous studies (Glikson 1984) to be isotopically ‘heavy’.

Londry and De Marais (2003) by conducting experiments with a number of sulphate reducing bacteria found that a variety of conditions such as rate of acetate

Fig. 2.5 Pure concentrate of humic acids (HA). HA in TEM appear as minute globules of medium electron density, often high electron density and electron opaqueness caused by metals absorbed by the HA. ×28,000

consumption of the bacteria effect the isotope composition of the Biomass/degraded OM. One species of sulphate reducers according to the same study resulted in C incorporated in the biomass being 8–9‰ isotopically heavier than the rest of the OM. The majority of microbial remains in the torbanites closely resemble methanogens found today in anaerobic environments. The other microbial forms observed in TEM of torbanites are bundles of filaments (Figs. 2.7; 2.9) of possible sulphate reducing bacterial affinity.

The amount of degradation in Botryococcal colonies is minimal; it mostly effects the periphery of the colonies (which is past oil generation) as indicated by the brown fluorescence in contrast to bright yellow of the unaffected algal parts. It has been noted by Cane and Albion (1971) that the role of micro-organisms especially anaerobic bacteria is of significance in alteration of the organic matter in the torbanites. The torbanite from Glen Davis revealed microstructures of spherical nature with numerous membranes reminiscent of cysts of *Methylobacter*, a methane utilizing bacterium, another factor potentially influencing carbon isotope composition. The cyst-like spherical bodies are 1–2 μ in size (Fig. 2.4). The microbial component in these torbanites may be responsible for the presence of monocarboxylic acids mentioned by Allan et al. (1979). Djuricic et al. (1978) analysed the bitumen and 'kerogen' (OM) fractions of an Australian torbanite and found the 'kerogen'/OM composition to be consistent with the results reported by Cane and

Fig. 2.6 TEM of torbanite: compressed algal mats (light grey) not always retaining their 'compressed mat' structure, which is often ruptured. Lipid bodies (LB/HC) very light colour, electron transparent clearly distinguishable. Harder OM is more electron dense and appears darker. An earlier study using TEM (Glikson et al. 1989) showed the changes occurring to the 'lipid bodies' during heating. Arrow points to stringers of possible sulphate reducing bacteria. Mag ×22,500. Scratch marks from sectioning knife seen perpendicular to the OM

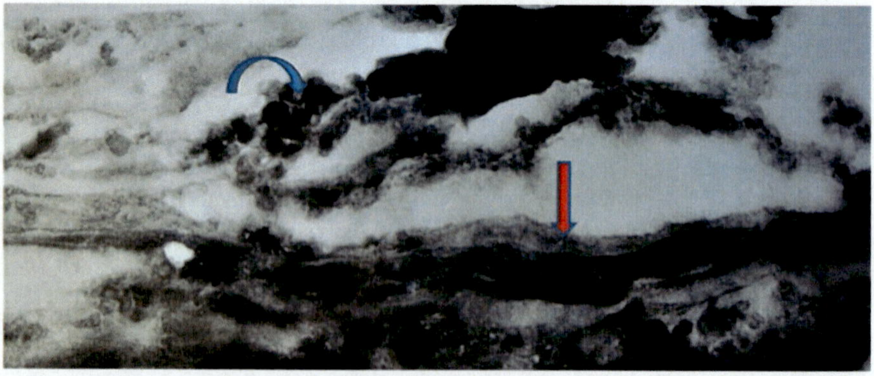

Fig. 2.7 Stringers of bacteria. Blue arrow points to pyrite and humic acids, red arrow points to bundles of filamenous structures. ×90,000 mag

Table 2.2 C-isotope composition of torbanites, Coorongite and *Botryococcus*

Torbanites	Sample	Fraction	δ^{13}C PPT
Marangaroo	TM-1	Total OM	−18.1
Alpha	TA-1	Total OM	−19.1
Cessnock	TK-1	Total OM	−19.5
Carnarvon Creek	TC-1	Total OM	−24.6
		Humic acids	−25.6
Coorongite (CSIRO No. 25233) Coorong South Australia	COO-1	Total OM	−17.8
Rubbery *Botryococcus* accumulations from Poorrarecup Lagoon, Western Australia	COO-2	Total OM	−9.8

Albion (1971) namely being derived from the main metabolites of *Botryococcus braunii*. The authors suggested Beta-carotene of *Botryococcus* orange resting state as a source, which may account for the aromatic hydrocarbons, while some of the aliphatic hydrocarbons may have their origin in microbial lipids.

An intriguing and interesting experimental study was conducted by Iyer et al. (2012) linking the morpho-physiological characteristics of *Botryococcus* to processes controlled by the genetic make-up of the alga. Thus, changes in the colour and shape of the cells have been shown to be expressions of changes in the bio-chemical stage of the alga, such as oil secreting or maturing and dying. Furthermore, the study demonstrated experimentally transformation of *Botryococcus* colonies to Coorongite when placed from fresh water into saline solution. Coorongite therefore is not merely a stage in the formation of torbanite as previously assumed. It is a condition of the alga *Botryococcus*in response to the environment. Coorongite forms today in brackish lagoons in South Australia. Torbanite formed in fresh water bodies.

Coorongite has been assigned to deposits of rubbery nature composed solely by the alga *Botryococcus braunii*. It has been forming in the Coorong lagoon, South Australia. Recent rubbery accumulations of *Botryococcus*—dominated deposits from other parts of Australia as well as blooms of *Botryococcus* from fresh water reservoirs have all been assigned to Coorongite (Wake and Hillen 1981). The present study shows that Coorongite may vary in its general composition according to its depositional environment and aerial exposure time. A sample from the Coorong lagoon observed in transmitted and reflected light has a considerable input from minerals dominated by diatom frustules, often cemented into clusters by amorphous organic matter which contains framboidal pyrite. TEM observations of the organic matter concentrate revealed humic acids and fossil bacteria. These observations as well as earlier ones (Glikson 1983) point to degradation prior to burial under anaerobic conditions. The abundance of pyrite is consistent with deposition in an anoxic environment.

TEM observations of rubbery *Botryococcus* accumulations from Poorrarecup lagoon in Western Australia (Fig. 2.11) displayed homogenous organic matter closely resembling ultra-thin sections of *Botryococcus* remains in torbanites. Some alteration, due to diagenesis and compaction is apparent both in the recent *Botryoccocus*

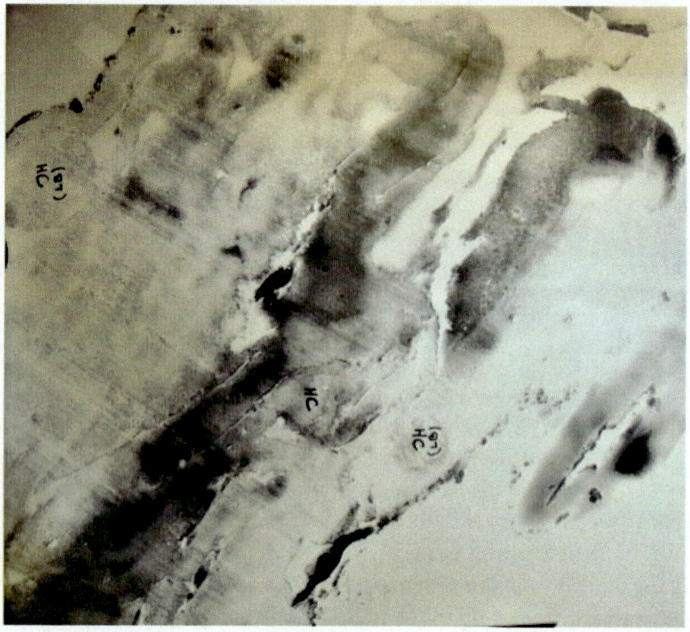

Fig. 2.8 Torbanite, pure algal material with lipid/hydrocarbon globules (LB, HC); ×26,000 mag

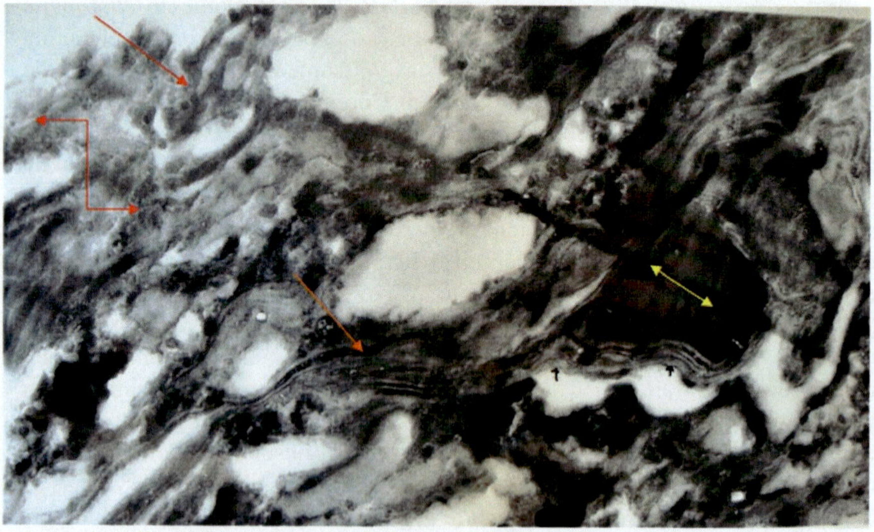

Fig. 2.9 Bundles of filamentous structures (filamentous possibly sulphate reducing bacteria) in torbanite. Algal material (electron transparent HC), and degraded OM associated with humic acids (red arrow). Inertinite (produced from heat effected alginate, yellow arrow). White, irregular shaped voids (V) are spaces where minerals have been dissolved during OM concentration procedure. × 90,000 mag

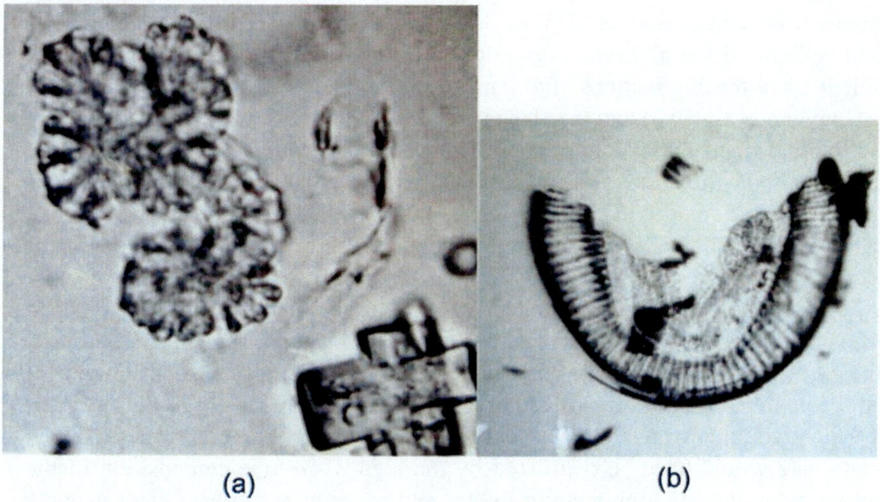

Fig. 2.10 Algal colonies in Coorongite, transmitted light microscopy, **a** gypsum crystal in lower right corner. **b** Diatom remains; Taken at ×100 mag (compare to Fig. 2.1)

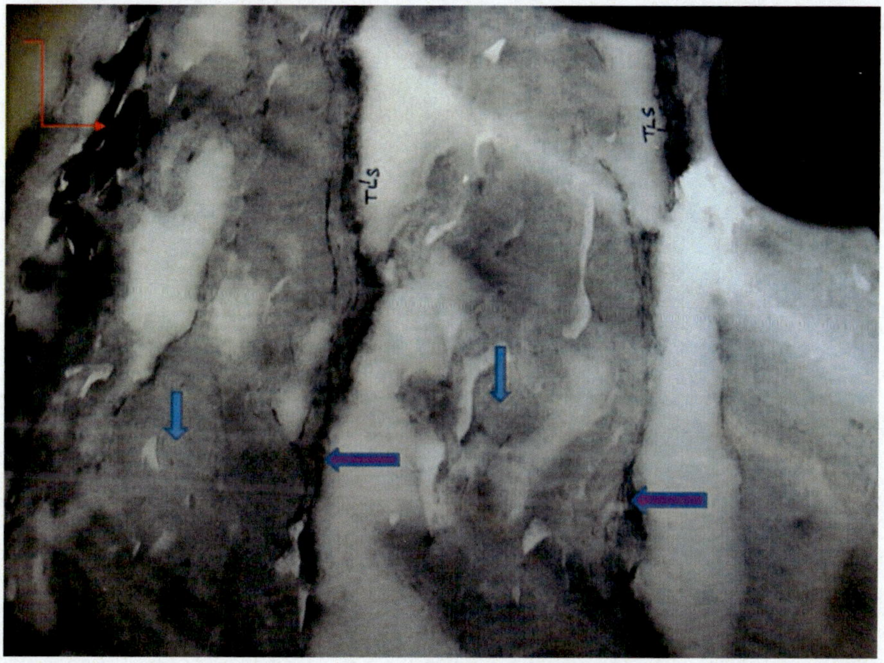

Fig. 2.11 Coorongite/layers of *Botryococcus*. Tri-lamellar structures (TLS) along every compressed individual alga. Outlines of lipid bodies can be distinguished (blue arrows); TLS along algal colony boundaries (purple arrows); metal concentrations (red arrow). ×52,000

accumulations of Coorongite algal remains, but more so in Torbanite algae. Individual cells are discernible in coorongite but rarely in torbanite. Spherical bodies visible in the ultra-thin sections of torbanites are lipid bodies; the hydrocarbon globules (Largeau et al. 1980) that the alga is known to produce. These hydrocarbons are toxic to other algae and plankton and explains why *Botryococcus* is the sole algal component in the torbanites and generally in fresh water lakes in the geological past.

The hydrocarbon or lipid bodies (Figs. 2.6; 2.8; 2.11) were compared by Largeau et al. (1980) to cuticular waxes of higher plants which explains the high degree of preservation of these structures in the Permian torbanites. The relatively isotopically 'heavy' values obtained for Coorongite have been discussed by Sacket and Thompson (1963) and explained as values typical of diatom dominated organic matter. Beardal et al. (1976) showed that diatoms when under conditions of high algal densities utilise bicarbonate in photosynthesis instead of CO_2. The coorongite is a high-density algal mat which would have restricted access to CO_2. The use of carbon from dissolved bi-carbonate would have been utilized by the alga. C-isotope composition of the Permian torbanites are significantly lighter and more in accordance with isotopic composition of plankton generally which points to a fundamental difference between torbanite *Botryococcus* and Coorongite *Botryococcus*. The Queensland torbanite which is closely associated with cannel coal has the isotope composition of coal (Table 2.3).

Sulphur increases in Torbanites almost exponentially with increase in C (Fig. 2.12). Generally S content is not as high as in Tasmanite which seems to contain very high Sulphur. A recent study by Hackley et al. (2017) of Tasmanite from the Upper Devonian of Ohio reported high S associated with organic C.

The S in Kukersite seems slightly higher than in Torbanites. The close association of Sulphur with organic Carbon is an important factor that needs to be considered when processing oil shales. Special treatment is required to remove the sulphur.

Extensive studies of the living alga *Botryococcus* conducted over the years, and more recently experiments growing the alga under varied conditions and analysing the changes (Iyer et al. 2012) that the alga undergoes, it became evident that Coorongite is a state of the alga *Botryococcus*, the result of environmental changes such as the increase in salinity. The same study demonstrated the transformation of *Botryococcus* to Coorongite when placed in saline solution. Interestingly the Coorong lagoon in

Table 2.3 Major elements

Sample	C	H	O	N	S	Ash	H/C	O/C
Coorongite	16.56	3.41	16.86	0.40	1.40	52.15	–	0.77
Torbanite								
TA-1	63.94	6.62	20.92	0.79	0.49	3.18	1.23	0.25
TA-2	75.06	10.09	6.97	0.38	1.88	5.00	1.60	0.07
TC-2	78.65	9.04	5.87	0.16	3.73	3.93	1.37	0.06
TC-1	45.91	4.92	9.72	0.60	0.32	37.93	1.28	0.16

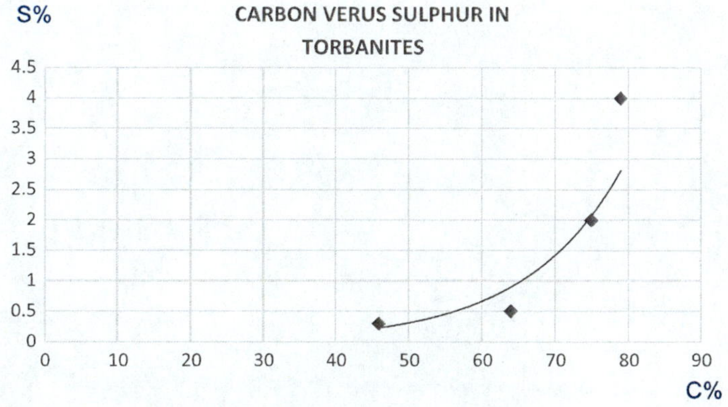

Fig. 2.12 Relation between carbon and sulphur in torbanites

South Australia for example is brackish, not a fresh water environment. The *Botryococcus* in torbanites on the other hand formed within fresh water inundation of the Permian coal forming swamps (Glikson-Simpson 2020).

Recently a number of studies of *Botryococcus* target the specific hydrocarbon biosynthesis gene of this alga (Browne et al. 2017; Hirano et al. 2019) which triggers the natural hydrocarbon production reported as 25–75% of the algal dry weight. *Botryococcus* algae on the basis of the above studies are seen as "one of the most promising candidates for biofuel production".

2.2 Tasmanite

The Tasmanite deposits of the Upper Carboniferous in the Mersey River sediments in Tasmania were mined from 1912 until the late 1940s. It was reported that 1.6 million liters oil shale was produced from these deposits over the years of mine operations (McLeod 2002).

Tasmanite owes its high oil potential to its precursor organism, *Tasmanites*, which is present in the form a cyst stage of a fossil alga with affinity to the modern green planktonic algae the Prasnophytales. The algae of this group form motile cells within an envelope of predominantly lipidic composition and thus are preserved in the geological record. The lipidic nature of *Tasmanites* cysts is evident from observations in light microscopy, fluorescence mode. The wall of the cyst fluoresces yellow in immature sediments and turns brown with advanced maturation similar to exines of pollen and spores. The motile cells which are not preserved in sediments are released from the cyst through canals visible in the fossilised cysts; Fluorescing *Tasmanites* cysts are illustrated and detailed in Vigran et al. (2008). In TEM the canals in the cyst wall are clearly visible (Fig. 2.13), although compressed over time. The cysts of

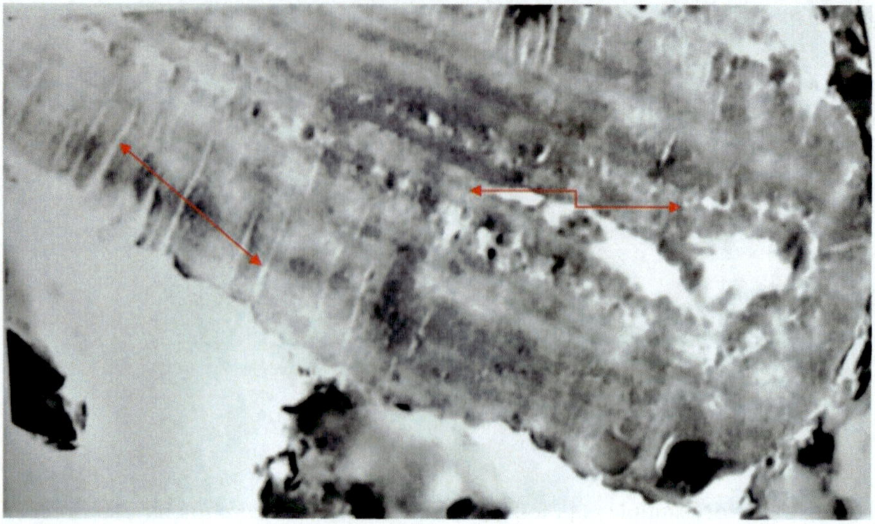

Fig. 2.13 TEM section of a part of *Tasmanites* cyst; showing canals (straight arrow) and spongy inner layer of the cyst in stages of disintegration (bent arrow). ×72,000

Tasmanites settle to the bottom of the water body after the motile cells are released. They are lipid-rich and resistant to degradation, flattened and preserved in the anoxic sediments.

Carbon Isotope Composition

$\delta^{13}C$ values for pure Tasmanite reported by Revill et al. (1994) are significantly 'heavy' values namely −13.5 to −11.7‰, similar to *Botryococus* isotopic values in Coorongite which is associated with diatoms. Similarly, there is an association of the alga with diatoms in Tasmanite. Revill et al. (1994) explain the 'heavy' values as typical of diatoms and CO_2 deprivation in 'sea ice' conditions. The Upper Carboniferous was heading towards the end of the glaciation of Gondwana. However, Coorongite is forming today in lagoons in temperate climate regions of South Australia. Turner and Fritz (2011) explain enrichment in 13C of interstitial waters due to their being saturated with calcium carbonate. However, the algae need light to photosynthesise which would be happening in the upper water column. It is suggested here that the tightly layered algal mats inhibited CO_2 circulation and C from sea water carbonate has been substituted.

TEM of *Tasmanites* (Fig. 2.13) shows numerous 'channels/canals' through the cyst of the alga which were the exit channels for the flagellate stage mini-spores

Table 2.4 Major element composition of tasmanite

C	H	O	N	S	Ash	H/C	O/C
72.97	9.52	9.40	0.81	8.91	6.69	1.55	0.10

Note the high S associated with high organic carbon

leaving the cyst. A two layered cyst wall is observed: A main thick layer of the cyst containing the 'channels' and a thin inner semi-detached partly degraded layer lining the central cavity of the spore cyst.

Vigran et al. (2008) described *Tasmanites* from the middle Triassic of Svalbard and the Barent Shelf. The same study suggests that *Tasmanites* is a major source of the hydrocarbons in sediments of that area. The Prasinophycean group of algae which *Tasmanites* belongs to, occur today in the Mediterranean and the Bay of Biscay and form blooms at times.

2.3 Kukersite

Another high oil yielding algal sourced oil shale deposit. The main difference between the kukersite deposits and the torbanites and tasmanite oil shales is the extensive size of the kukersite oil shales deposit compared to small lenses of torbanites and tasmanite. Kukersite oil shales have been mined for over 100 years and are the main supplier of energy to power generation in Estonia. The oil shale is mined above ground as well as in underground room and pillar mines. The kukersite oil shales are also used as a direct combustion fuel in electricity generation, similar to the use of coal.

Kukersite is the name given to oil shales of Middle Ordovician age in a basin which covers 50,000 km^2 in northern Estonia and extends eastward into Russia (Dyni 2006). The main oil shale formation contains up to 50 organic-rich kukersite beds alternating with argillaceous limestone (Bauert 1994). The organic-rich beds 10–40 cm thick in the deposit are principally made up of the alga *Gloeocapsomorpha prisca.* This alga is intertidal, to very shallow subtidal, marine colonial mat-forming cyanobacterium. A detailed and comprehensive study by Foster et al. (1989) highlighted the close resemblance of *G. prisca* to the living cyanobacterium *Entophysalis major*, and the fossil *E. belcherensis.*

Golubic and Abed (2010) in detailed studies of living cyano-microbial mats in Abu Dhabi demonstrated the precise favoured environments and conditions for *Entophysalis* growth and mat forming. The intensively studied stromatolites of Shark bay and numerous other lagoonal lakes in Western Australia are well documented and explain the seasonal effect on the growth and composition of cyanobacteria in stromatolites. Of particular relevance to kukersite is the fact that these stromatolites are so rich in calcium carbonate: It was pointed out by McNamara (1997) regarding stromatolites that: "The link between all the living Western Australian stromatolites, irrespective of the salinity of the water in which they grow, is that they develop in environments rich in calcium carbonate; either lime-rich sand or limestone. Groundwater that has passed through such lime-rich substrates may be the source for the stromatolites' structure." Such as those seen today at Lake Clifton, Western Australia (Fig. 2.14).

Fig. 2.14 Lake Clifton stromatolites (photo by McNamara 1992)

Table 2.5 Major elements % in kukersite

	C	H	O	N	S	Ash
In whole rock (this study)	66.81	8.04	12.85	0.43	2.46	5.8
In OM concentrate (Bauert)	76–77	9.4–9.9	9.0–11.0	0.2–0.5	1.2–2.0	H/C = 1.48

Exceptionally high content of calcium carbonate characterises stromatolites as well as kukersite. The oil shale seams are interbedded with Limestone which is used in Estonia in the production of cement (Table 2.5).

Detailed information on mining, production and environmental issues of kukersite are outlined in Bauert and Soesoo (2007). Some of these will be discussed in Chap. 7.

2.4 Green River Oil Shale

This oil shale, like kukersite, is another deposit with high H/C ratio that owes its high oil yield to algal accumulations. The Green River oil shale is the world's largest known oil shale deposit (Shanks et al. 1976). It extends over 65,000 km^2 through the states of Wyoming, Utah and Colorado which were occupied in the Lower and

Middle Eocene by an extensive lake system for over 10 million years, during a time of warm temperate to sub-tropical climate (Dyni 2006).

During dry periods these lakes shrank, and upon renewed wet phase transgressed again, resulting in deposition of deep-water organic-rich shales. The ensuing deposits include lacustrine, littoral and evaporitic types, depending on the lake water level fluctuations. The formation thus represents a complex cycle of varied lithologies, ranging from mudstone to limestone to oil shale. Each lithological unit varies in thickness and is intercalated with sandstones in the upper part of the Formation (Shanks et al. 1976). Bradley (1970) in detailed studies of the Green River sediments explained the formations as being the result of lake deposits, hydrologically closed and permanently stratified. Bradley favoured cyanobacteria as the dominant source of OM. However, a detailed isotopic geochemical study by Hayes et al. (1989) showed that many sources contributed to the Green River organic-rich oil shales. Cyanobacteria, according to their study contributed a small amount, so did methanotrophic microbes, while the bulk of OM was sourced from '*highly aliphatic bio-polymers of eukaryotic origin*'. The same study established geochemically that the organic matter was 'directly inherited from lipids'. TEM of the present study confirms visually Hayes et al.'s analytical results by showing the OM as having characteristics of non-degraded algal remains, possibly *Botryococcus.* H/C ratios are the same as for high oil yield *Botryococcus* dominated oil shales.

As the present study is focused on the Australian oil shales, only a limited number of samples were studied, all from Lake Gosiute Mine in Wyoming and are reported here. TEM of Green River OM displayed characteristic algal OM with pockets of humic acid concentrations associated with metals, mainly Cr-Ti-Fe. Cr was shown from Trace Element analysis to be abnormally high, and so did Ti. However only iron (pyrite) showed strong association with the humic acids, while Cr and Ti were expressed by small peaks only, in EDS of organic concentrate. This implies that the high concentrations of Cr and Ti were mainly in the inorganic fraction. The source of Ti is in volcanic rocks while Cr is sourced mainly from mafic and ultra-mafic rocks. A study by Poznanovic-Spahic et al. (2018) showed Cr to be a lithophile element which can substitute Al and Fe in minerals, which may have a bearing on its high concentration in the mineral fraction.

In light microscopy, using UV light/blue filter excitation, the predominant OM in the Green River exhibits bright yellow fluorescence while some OM of higher maturation fluoresces orange-brown, highlighting the oil potential. The mode of appearance of this OM in TEM observation is unusual: It is inter-connected as a one continuous organic mass (Figs. 2.15; 2.16). Flow lines are recognisable where electron dense particles (metals) must have been carried through the organic mass, marking 'boundaries' between algal walls (Figs. 2.17; 2.18). Brines percolating through these deposits may have brought about chemical changes which are responsible for the obliteration of most of the primary structures of the precursor organism. Surdam and Wolfbauer (1975) in their study of the Green River Formation in Wyoming, and in particular Lake Gosiute explain the deposit in terms of a playa-lake complex. They found that the pattern of lithological facies and mineral concentrations are identical with today's playa-lake settings. If that was the case, comparison can be made

Fig. 2.15 TEM of Green River OM showing the predominant algal material. Gaps and empty spaces represent carbonates and silicates that have been dissolved during OM concentration procedure. The black, highly electron dense particles represent metals or high atomic number minerals. Note the algal material mostly connected. Taken at ×4500

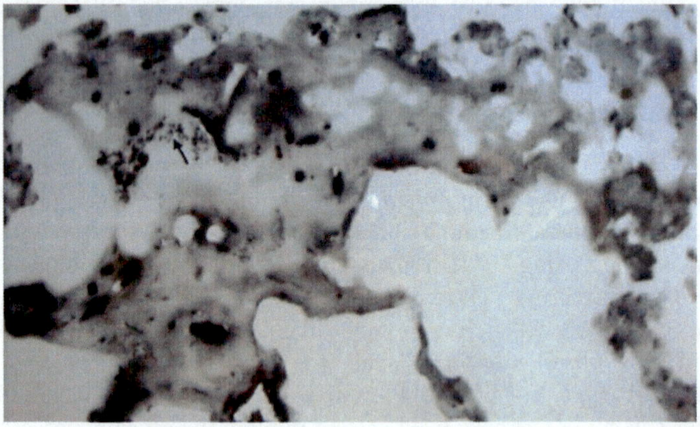

Fig. 2.16 Detail of Fig. 2.15

Fig. 2.17 Detail of an area in the algal material showing concentrations of metals/high atomic number minerals (black, electron opaque particles, yellow arrow) filling voids in the algal mass, and boundaries along preserved algal walls marked by very fine aligned mineral grains (blue and small black arrows). ×72,000

with the study of microbial mats-/ stromatolites in Abu Dhabi (Golobic and Abed 2010) and the stromatolites forming today in similar settings in Western Australia. However, the detailed isotope—geochemical study of Hayes et al. (1989) clearly demonstrates that cyanobacteria were only a minor contributor to the Green River OM. The C-isotope results of the present study are in agreement with Hayes (1993) results. The TEM observations of the present study show typical green algal OM such as *Botryococcus* or *Pediastrum* as predominant.

C-isotope compositions of the OM in the Green River obtained in this study (Table 2.6) are $\delta C^{13} = -30.30‰$ for the OM less the Humic acids, and −29.9‰ for extracted lipids. Collister et al (1993) obtained similar values for $\delta C^{13} = -28$ to −32‰. Degens et al. (1968) however reported chloroform extractable lipids of plankton in the Pacific had a mean isotopic value of −29.8‰, identical to the isotope composition obtained in the present study.

EDS analysis showed the metals were predominantly Iron sulphides (pyrite) and some Phosphorus. Only traces of Ti and Cr were detected in the OM fraction, as the bulk of these metals is associated with the inorganic, mineral fraction. Occasionally cell walls in the algal matter are recognisable (arrow). The very fine lines (arrows in Fig. 2.17) of tiny electron dense particles are reminiscent of well documented

Fig. 2.18 TEM showing metals (electron opaque particles; thick blue arrows) in algal remains along flow-lines. Humic acids (thin red arrow) absorbed metals (opaque, black areas). ×117,000

Table 2.6 Major elements %, trace elements ppm, and 13C‰

C	H	O	N	S	Ash	H/C	O/C
68.57	8.79	8.94	0.93	3.54	8.70	1.53	0.10
Trace elements							
Ti ppm	V ppm			Cr ppm			
6516	23			4010			
$\delta^{13}C$ composition							
Fraction				$\delta^{13}C$			
Chloroform extracted lipids				−29.9‰			
OM less humic acids				−30.3‰			

lamellae (TLS) in *Botryococcus* algae. Likewise, the electron transparency of the bulk material is identical to that of the green algae *Botryococcus and Pediastrum.*

References

Allan, J., Bjoroy, M., & Douglas, A. G. (1979). A geochemical study of the exinite group maceral alginite selected from three Permo-Carbonifetous torbanites. In A. G. Douglas & J. R. (Eds.), *Ad. Organic Geochemistry* (pp. 599–618).

Bauert, H. (1994). Kukersite-An oil shale of Ordovician age: Origin, occurrence and geochemistry. In *Confernce Kukersite-An Oil Shale of Ordovician age, OSTI. Gov* (10 pp.).

Bauert, H., & Soesoo, A. (2007). Kukersite—An oil shale of Ordovician age: Origin, occurrence, and geochemistry. In *Osti.Gov./Conference*.

Beardal, J., Glover, H. E., & Morris, I. (1976). The path of carbon in the photosynthesis of marine phytoplankton. *Journal of Phycology, 12,* 409–417.

Bradley, W. H. (1970). Green River oil shale—Concept of origin extended. *Geological Society of America Bulletin, 81,* 985–1000.

Browne, R. B., Jenkins, J., Schmutz, J., Shu, S., Barry, K., Grimwood, J., … & Devarenne, T. (2017). Nuclear Genome sequence of the liquid hydrocarbon – accumulating in microalga Botryococcus braunii Race A. *Genome Announcements, 20*(16), 5.

Cane, R. F., & Albion, P.R. (1971). The phytochemical history of torbanites. *Journal and Proceedings of the Royal Society of New South Wales, 104,* 31–37.

Collister, J., Summons, R. E., Litchfouse, E., & Hayes, J. (1993). An isotopic biochemical study of the Green River oil shale. *Organic Geochemistry, 19*(1–3), 265–276.

Connah, T. H. (1964). Torbanite deposit—Alpha, Central Queensland. *Report of Geological Survey of Queensland, 3,* 1–5.

Degens, E. T., Guillard, R. L., Sackett, W. M., & Hellebust, J. A. (1968). Metabolic fractionation of carbon isotopes in marine plankton. *Deep Sea Research, 15,* 1–9.

Derenne, S. (1997). Chemical structure of the organic matter in a Pliocene maar-type shale: Implicated *Botryococcus* race strains and formation pathways. *Geochimica et Cosmochimica Acta, 61,* 1879–1889.

Derenne, S., Largeau, C., Casadevall, E., & Sellier, N. (1990). Direct relationship between the resistant biopolymer and the tetraterpenic hydrocarbon in lycopadiene race of *Botryococcus braunii. Phytochemistry, 29*(7), 2187–2192.

Djuricic, M., Vitorovic, D., Hertz, D., Anderson, B., & Biemann, K. (1978). Relationship between the bitumen and kerogen of an Australian torbanite. *Organic Geochemistry, 1,* 79–84.

Dulhunty, J. A. (1941). Origin of New South Wales torbanites. *Proceedings of the Linnean Society of New South Wales, 66,* 257.

Dulhunty, J. A. (1943). 1944: Origin of New South Wales Torbanites. *Proceedings of the Linnean Society of New South Wales, 68, 69.*

Dyni, J. R. (2006). *Geology & resources of some world oil shales deposits* (USGS Report, 41 pp).

Foster, C. B., Reed, J. D., & Wicander, R. (1989). *Gloeocapsomorpha prisca* Zalessky, 1917: A new study: Part I: Taxonomy, geochemistry, and paleoecology. *Geobios, 22*(6), 735–759.

Glikson, M. (1983). Microbiological precursors of coorongite and torbanite and the role of microbial degradation in the formation of kerogen. *Organic Geochemistry, 4,* 161–172.

Glikson, M. (1984). Further studies on torbanites and Coorongite using transmission electron microscopy and C-isotope analysis. *Organic Geochemistry, 7*(2), 151–160.

Glikson-Simpson, M. (2020). *Coal-Window to Past Climate and Vegetation* (180 pp.). Springer.

Glikson, M., Lindsay, K., & Saxby, J. (1989). *Botryococcus*—A planktonic green alga, the source of petroleum through the ages: Transmission electron microscopical studies of oil shales and petroleum source rocks. *Organic Geochemistry, 14*(6), 595–608.

Glikson, M., Taylor, D., & Morris, D. (1992). Lower Palaeozoic and Precambrian petroleum source rocks and the coalification path of alginate. *Organic Geochemistry, 18*(6), 881–897.

Golobic, S., & Abed, R. M. M. (2010). Entophysalis mats as environmental regulators. In J. Seckbach & A. Oren (Eds.), *Microbial mats: Modern and ancient microorganisms in stratified systems* (pp. 237–251). Springer.

Hackley, P. C., Walters, C. C., Kelemen, S. R., Mastalerz, M., & Lowers, H. A. (2017). Organic petrology and micro-spectroscopy of *Tasmanites* microfossils: Application to kerogen transformations in the early oil window. *Organic Geochemistry, 117,* 23–41.

Hayes, J. M. (1993). Factors controlling ^{13}C contents of sedimentary compounds-principles and evidence. *Marine Geology, 113,* 111–125.
Hayes, J. M., Popp, B. N., Takigiku, R., & Johnson, M. W. (1989). An isotopic study of biogeochemical relationships between carbonates and organic carbon in Greenhorn Formation. *Geochimica et Cosmochimica Acta, 53,* 2961–2972.
Hedges, J. I. (1978). The formation and clay mineral reactions of melanoidins. *Geochimica et Cosmochimica Acta, 42,* 69–76.
Hirano, K., Hara, T., Ardianor, Agung-Nugroho, A., Segah, H., Takayama, N., Sulmin, G., … & Kawamura, K. (2019). Detection of oil-producing microalga Botryococcus braunii in natural freshwater environmentsm by targetting the hydrocarbon biosynthesis geme SSL-3. *Scientific Reports, 18*(9), 1674.
Hobson, P. N., Bousfield, S., & Summers, R. (1974). Anaerobic digestion Hobson P.N., Baker B.L, Peake E. 174: Geochemistry of Porphyrins. In: Nagy & Colombo (Eds.), *Fundamental Aspects of Petroleum Geochemistry.* Elsevier, Amsterdam.
Iyer, G., Menon, S., & Gupte, Y. (2012). Morpho-physiological characteristic of *Botryococcus braunii* (Kutzig 1849) and its oil production from species isolated from Thane, Maharashtra, India. *Asian Journal of Microbiology, Biotechnology and Environmental Sciences, 14*(4), 523–535.
Largeau, C., Casadevall, E., & Berkaloft, C. (1980). The biosynthesis of long-chain hydrocarbons in the green alga *Botryococcus braunii. Phytochemistry, 19,* 1081–1085.
Londry, K. L., & De Marais, D. J. (2003). Stable carbon isotope fractionation by sulfate-reducing bacteria. *Applied and Environment Microbiology, 69*(5), 2942–2949.
Madre, D. A. (1987). The geology of the alpha oil shales deposit. *Fuel, 66,* 309–313.
McLeod, R. (2002). The tasmanite oil shale resource Latrobe-Railton area, Tasmania. A review.
McNamara, K. J. (1992). *Stromatolites* (28 pp.). Western Australian Museum Publication.
McNamara, K. J. (1997). *Stromatolites* (28 pp.). Western Australian Museum Publication.
Metzger, P., & Largeau, C. (2005). *Botryococcus braunii*: A rich source for hydrocarbon and related ether lipids. *Applied Microbiology and Biotechnology, 66,* 486–496.
Poznanovic-Spahic, M., Sakan, S., & Cvetkovicz-Tcic, P. (2018). Environmental risk and origin of heavy metals. *Environmental Monitoring and Assessment, 190*(4), 190–208.
Reddy, C. A., Bryant, M. P., & Wolin, M. J. (1972). Characteristics of 'S' organism isolated from *Methanobacillus omelianskii. Journal of Bacteriology, 109,* 539–545.
Revill, A. T., Volkman, J. K., O'Leary, T., Summons, R. E., Boreham, C. J., Banks, M. R., & Denver, K. (1994). Hydrocarbon biomarkers, thermal maturity, and depositional setting of Tasmania, Australia. *Geochimica et Cosmochimica Acta, 58,* 3803–3822.
Sacket, W. M., & Thompson, R. R. (1963). Isotopic organic carbon composition of recent continental derived clastic sediments of eastern gulf coast, Gulf of Mexico. *AAPG Bulletin, 47,* 525–531.
Saxby, J. D., Bennett, J. F., Corcoran, D., & Lambert, D. E. (1986). Petroleum generation: Simulation over six years of hydrocarbon formation from torbanite and brown coal in a subsiding basin. *Organic Geochemistry, 9*(2), 69–81.
Shanks, W. C., Seyfried, W. E., Meyer, W. C., & O'Neill, T. J. (1976). Mineralogy of oil shale. In T. F. Yen & G. V. Chillingarian (Eds.), *Oil shale development in petroleum science* (Vol. 5, pp. 81–102). Amsterdam: Elsevier.
Surdam, R. C., & Wolfbauer, C. A. (1975). Green River Formation, Wyoming: A playa-lake complex. *Geological Society of America Bulletin, 86,* 335–345.
Turner, J. V., & Fritz, P. (2011). Enriched ^{13}C composition of interstitial water in sediment of fresh water lake. *Organic Geochemistry, 1993,* 265–276.
Vigran, J. O., Mork, A., Forsberg, A. W., Weiss, H. M., & Weitschat, W. (2008). *Tasmanites* algae—Contributors to the Middle Triassic hydrocarbon source rocks of Svalbard and the Barents Shelf. *Polar Research, 27,* 360–371.
Wake, L. V., & Hillen, L. W. (1981). Nature and hydrocarbon content of blooms of the alga *Botryococcus braunii* occurring in Australian freshwater lakes. *Australian Journal of Marine and Freshwater Research, 32,* 353–367.
Zeikus, J. E. (1977). The biology of methanogenic bacteria. *Bacteriological Reviews, 41,* 514–541.

Chapter 3
Camooweal Oil Shales

Abstract The combination of various methodologies applied to Cambrian and Lower Palaeozoic oil shales from Georgina Basin, Australia are used to elucidate on their composition and mode of appearance. The combined use of light microscopy and Transmission electron microscopy as well as elemental analyses and C-isotope composition aided in elucidating the nature of some of the organic matter. The latter lacks any well-defined structures when observed in light microscopy that can be directly identified with an organism. Thus, TEM is the only means to observe any structured organic entities. TEM showed methanogenic coccoid microbial remains, cyanobacterial remains and humic acids. The activity of methanogenic bacteria may explain the very negative isotope composition of some of the humic acids since the humic acids are degradation products of the original organic matter. Reflectance (Ro) indicators of vitrinite-like organic matter helped to reconstruct the thermal history of the basin and elucidate the presence of several Ro populations of organic matter. Hydrothermal effect on organic matter is pronounced in the southern and south-western parts of the basin where oil and gas generated as a result of several hydrothermal surges along sets of major fault systems. These effects were not observed in the organic matter in the northern parts of the basin, namely the oil shales.

3.1 The Middle Cambrian Camooweal Oil Shales

The Middle Cambrian Camooweal oil shales form a part of the sediments of the Georgina Basin—one of the largest depositional basins in Australia, that covers over 330,000 km^2.

The Camooweal oil shales occur in the Middle Cambrian Current Bush Formation and are marked by a major transgression resulting in shallow seas over the greater part of the Basin. This environment with access to the open sea allowed nutrients supply and high organic productivity, while anoxic conditions in parts of the Basin enabled the preservation of the organic matter. The Camooweal oil shales form thin bands of OM within a calcareous sequence reminiscent of cyanobacterial mats. Depositional

M. Glikson-Simpson, *Oil Shales*, https://doi.org/10.1007/978-3-030-60675-6_3

environment may have been of limestone platforms in a shallow epicontinental sea as suggested by an earlier study (Cook 1982).

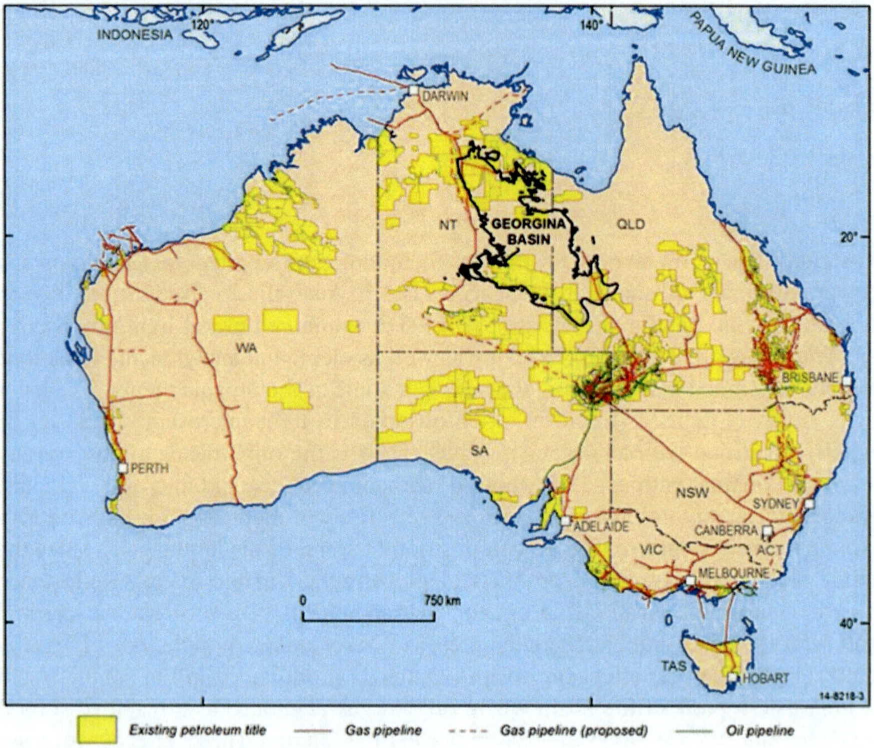

Location map. *Source* Geoscience Australia

The organic matter in the Camooweal oil shales is closely associated with carbonates as evident in reflected light microscopy of polished rocks (Figs. 3.2; 3.4) and the oil shales usually occur in horizons not more than 10 cm thick. The high carbonate content (Table 3.1) and type of OM are reminiscent of the *Entophysalis* mats forming today and described by Golubic and Abed (2010) in great detail from Abu Dhabi. Riding (1991) suggests that the Cambrian calcareous flora was dominated by cyanobacteria. Indirect evidence in support of cyanobacterial input is provided by GC-MS analyses (P. Philp, in Glikson et al. 1985). Cyanobacterial origins for the OM are suggested in the present study for the Camooweal oil shales; algal remains have been identified as the dominant OM in the southern part of the basin (Glikson et al. 1999), whereas cyanobacterial affinity is suggestive for the stringers and fine laminations of the OM in the northern part of the basin.

The organic matter in boreholes Mt Isa-1 and Camooweal in the northern part of the basin when classified on the base of its reflectance (Ro) was divided into three groups in an earlier study (Glikson et al. 1985) whereas, two of these appear to grade into one another, a third OM is distinct and separate. In the present study two groups are obvious in reflected light observations, (Figs. 3.2; 3.4), all parallel to bedding: A very low reflectance OM (Ro $\leq$ 0.4%) forming continuous fine bands and stringers separated by calcite is evident. This is the primary and predominant cyanobacterial organic material. In transmitted light most of the OM was amorphous with no definite structure of identifiable organisms except rare micro-structures resembling *Eontophysalis becherensis* (Fig. 3.1), such as have been described by Oehler (1978) from Middle Proterozoic sediments. The second type of OM clearly distinct in reflected light and recognisable has a smooth to faintly pitted surface and is of significantly higher Ro (>0.8%), usually accompanies the low reflectance laminae of type 1. This type is distinguishable in TEM as humic acids. A third type, rarely encountered OM of high Ro (Fig. 3.2b; 3.4b) with well-defined algal cell structures such as have been observed in earlier studies (Glikson et al. 1992). This OM, of relatively high Ro, although similar to type 2 at first glance, has visible surficial sculpture distinguishing it from Type 2 of same Ro. In TEM it is clearly discernible as thermally effected *Botryococcus* cells (Fig. 3.6) characteristically of high electron density (high reflectance in reflected light microscopy). This OM is rare and it is suggestive of re-worked algal remains from earlier deposition.

Interesting to note the variation in abundance of OM and minerals. Samples high in carbonate are lower in OM than samples with exceptionally high pyrite which have very high OM, indicating highly anoxic conditions enabling the preservation of more OM. Reflectance data of the Camooweal OM in an early study (Glikson et al. 1985) showed three classes of OM. However, there was a certain overlap between them.

Three types of OM were distinguished by Sandstrom (1982) in the Middle Cambrian Thorntonia carbonates of the Georgina Basin: 1. A predominant lamellar

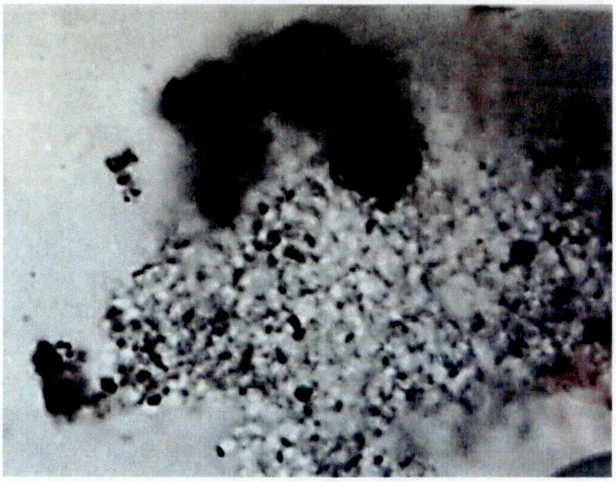

Fig. 3.1 Structures resembling *Eontophysalis* in Camooweal organic matter. Black spots and round particles are pyrite. Transmitted light. ×250

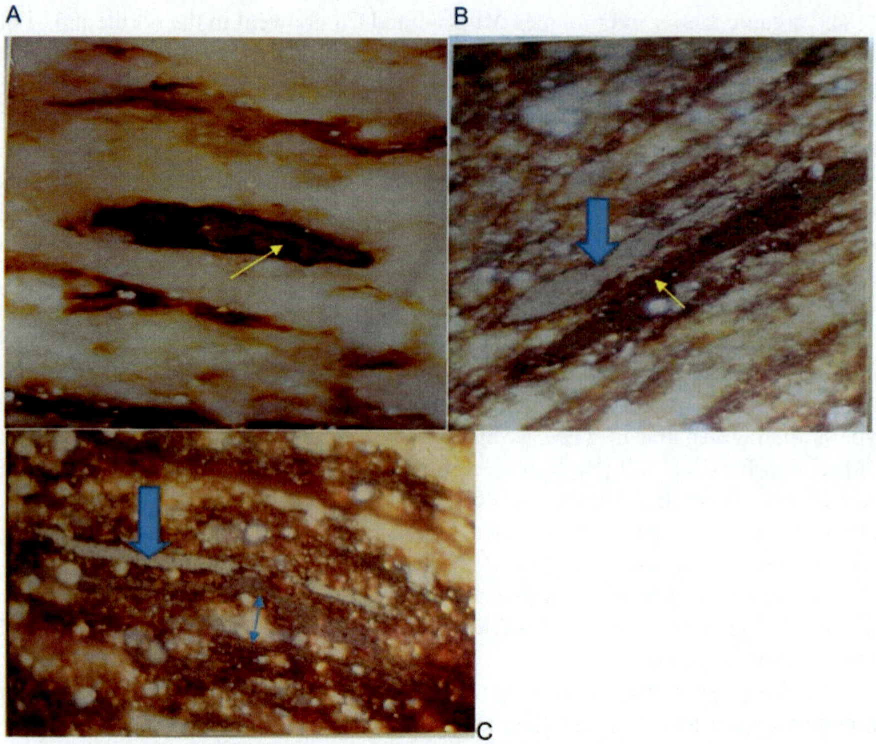

Fig. 3.2 **a–c** Reflected light microscopy of Camooweal oil shale: **a** very low Ro (yellow arrow) OM wedged within calcite. **b** and **c** Low Ro stringers of OM parallel to bedding (thin yellow arrows) and higher Ro OM, also parallel to bedding (block arrows). An intermediate Ro between very low dark stingers and high Ro is visible in C (thin blue arrow). White beds are calcite; bright white round particles are pyrite. Taken at ×100 mag

Table 3.1 Variation in abundance of the three main components of Camooweal oil shales

Sample	OM%	Pyrite%	Carbonate%
C-6	46	14	40
C-1	60	34	6
C-2	44	4	52

OM squeezed between and around mineral grains, dark grey/black in reflected incident light (low Ro) and displaying pale yellow to dull orange fluorescence in UV light (blue filter), which most probably corresponds to the extreme low reflectance dark grey lamellar predominant OM type in the middle Cambrian of the southern part of the basin, as no fluorescing OM has been observed in the samples from the northern part of the basin (e.g. Camooweal and Mt Isa borehole samples). Cyanobacteria do not fluoresce in UV light whereas green algae do. Algal remains have been reported from OM in south Georgina Basin (Glikson et al. 1992).

2. A significant fraction of humic acids has also been noted by Sandstrom (1982) in the Cambrian samples of his study; these would correspond to the type 2 higher reflectance OM of Camooweal oil shales in the present study.

Both type 1 OM and type 2 are consistently parallel to bedding. Type 1 namely, low Ro lamellar OM is the main organic component, the predominant cyanobacterial stringers in the oil shales. Some of type 2 display finely pitted texture when observed in reflected light, and is seen associated with type 1 (Fig. 3.2b, c; 3.4b). Pyrite (small white circular spots) abounds within type 1 OM. In the present study, most of type 2 OM was found to be extractable in NaOH and is therefore regarded as humic acids which are clearly distinguished in TEM (Fig. 3.6–blue arrow). The C-isotopic composition of this fraction was significantly 'lighter' than the rest of the OM with the exception of one sample. Extensive studies carried out on marine humates have shown that carbon in humic acids is isotopically close to or slightly 'heavier' than it is in plankton of the same environment. In view of these studies it is probable that the humic fraction of the Camooweal shales is a degradation product derived from other organisms. Other organisms seen associated with the OM in these deposits are presumed methanogenic bacteria identifiable in TEM (Fig. 3.5). These bacteria being responsible for the degradation of the OM (HA) preferentially metabolize the 'lighter' isotope and are therefore responsible for the ^{12}C enrichment of the HA fraction.

While in reflected light microscopy only two types of OM are clearly distinct in the Camooweal samples (Northern Basin); TEM allows the occasional separation of three types of OM (Fig. 3.6). 1. The low Ro possibly the primary OM (cyanobacterial) being electron transparent, forming bands/stringers along mineral grains (vacant spaces), 2. the medium electron dense globular structures (high Ro lenses in reflected light) being humic acids (HA) and 3. highly electron dense hard material in TEM reminiscent of algae subjected to thermal stress, corresponding to the rare occurrence of high Ro lenses of OM with surficial sculpture observed in reflected light (Table 3.2).

Carbon-isotope studies of OM in Australian Cambrian and Precambrian sediments carried out by McKirdy (1977) and Sandstrom (1982) obtained δ^{13} −30.5 to −32.6‰ for Cambrian in general and middle Cambrian OM from the Georgina Basin.

The present study of samples from the middle Cambrian Camooweal Oil shales yielded isotope compositions of −28.6 to −31.8‰, with one sample yielding a value of −36.4‰. A similar range in C isotope composition as obtained in the present study was obtained by Creveling et al. (2014) who report values of −28 to −30‰ for OM in the Middle Cambrian Thorntonia Limestone in the Georgina Basin. The humic acid fraction in the present study, when analysed separately was always slightly isotopically lighter in the majority of samples than the whole OM. Three anomalies of extreme isotope values of humic acids were obtained in three samples. The 'lighter' isotope values in the humic acid fraction point to the humates being a degradation product involving anaerobic bacteria. Their extremely negative isotope values suggest major involvement of sulphate reducing and methanogenic bacteria in the degradation process. Coccoid bacteria resembling methanogens adhering to the predominant OM are clearly visible in TEM (Fig. 3.5). Detailed studies of anaerobic

Table 3.2 C-isotope values of OM from the Lower Palaeozoic Precambrian oil shales

Name of well	Sample	Depth	Fraction analysed	$\delta^{13}C$‰
Lake Amadeus No. 1	LA-5	779′6″	OM less HA	−32.3
			Total OM	−36.4
			HA	−40.6
	LA-1	788′	Total OM	−29.1
Henbury No. 2	H2-5	770′10″	Total OM	−29.0
East Mereenie No. 1	EMI-2	3465′	OM less HA	−31.7
			HA	−31.8
	EM1-3	3470′	Total OM	−28.5
Mt Isa No. 1	C-1	117.43 m	Total OM	−28.6
	C-1		HA	−48.8
	C-2	83.55 m	OM less HA	−28.8
	C-2		HA	−31.2
	C-4	82.88 m	OM less HA	−27.0
	C-4		HA	−48.8
	C-6	95.52 m	Total OM	−30.9
	C-7	107.60 m	Total OM	−29.9
	C-8	114.50 m	OM less HA	−29.1
	C-9	117.40 m	OM less HA	−27.8
	C-9		HA	−31.6

OM = organic matter as a whole
HA = humic acids (separated from whole OM)

respiration in general and methanogenic bacterial activity specifically in sediments, demonstrated the process of fractionation and enrichment in $\delta^{12}C$ (Hayes 1993; Conrad et al. 2010a, b). The close composition of C-isotope values of most OM and its humic acid fraction suggests in situ degradation of the algae and associated microorganisms. The C-isotope composition of the HA anomaly in samples LA-1, C-1 and C-4 (Table 3.2) points to highly anoxic conditions at times and intense activity of methanogenic and other anaerobic bacteria.

Elemental analysis results show distinctive variation in organic carbon content; from as low as 18% C—to close to 70%. Ash which is an indicator of mineral content varies likewise. The higher the mineral concentration the lower the organic matter as can be clearly seen in reflected light micrographs. In Figs. 3.2 and 3.4 well over 50% minerals are visible and only 2–3 bands of OM.

Nitrogen is high in the Camooweal samples and is evidence of cyanobacteria being significant contributers to the OM. This association will be discussed in more detail in the chapter on the Toolebuc Formation oil shales.

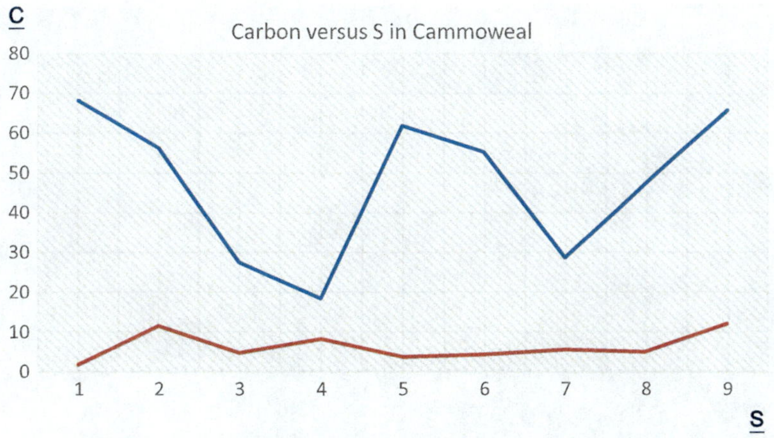

Fig. 3.3 Carbon and sulphur relationship

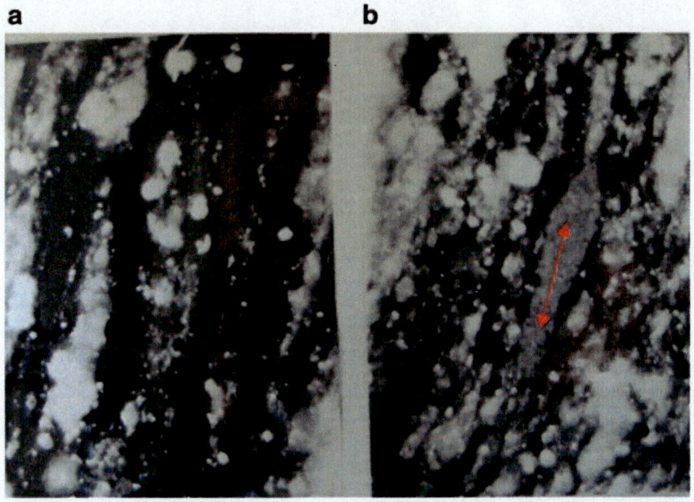

Fig. 3.4 **a)** Bands, stringers black and charcoal (slightly under the surface of the polished slab) of very low reflectance OM parallel to bedding, reflected light microscopy. **b)** High reflectance lense of OM (arrow) wedged between low Ro OM. Red arrow shows bright and dark patches at closer look are reminiscent of algal cells. Bright dots are pyrite; white areas are calcite. Taken at $\times 100$

Sulphur (Fig. 3.3) is high in some samples, it does not always show a positive correlation with carbon, only occasionally and is mostly tied up in pyrite. An exception are two samples that show a positive correlation, where high C in OM is also high in S.

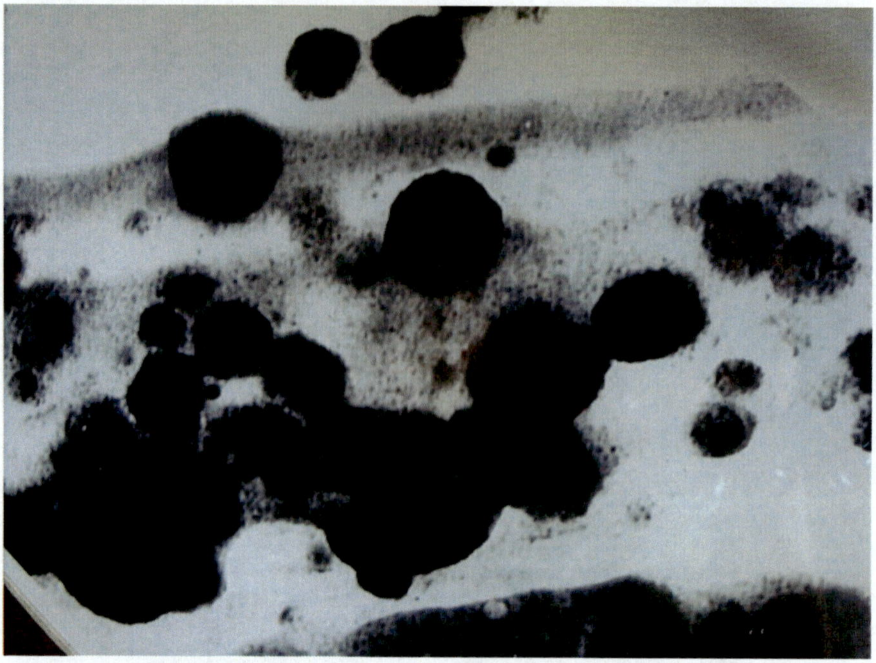

Fig. 3.5 Methanogenic bacteria embedded in the background organic matter. TEM ×120,000

Table 3.3 Major elements %

Samples ID	C	H	O	N	S	Ash	H/C
Mt Isa-1							
C-1	68.12	7.62	10.63	2.09	1.71	9.68	1.7
C-2	56.16	6.43	11.46	2.15	11.46	28.64	1.3
C-3	27.44	3.60	11.76	0.70	4.68	45.15	1.5
C-5	18.43	2.60	11.20	0.45	8.23	61.01	1.6
C-6	61.67	7.08	10.93	1.88	3.65	20.19	1.3
C-7	55.20	6.23	11.29	1.64	4.34	20.52	1.3
C-8	28.58	3.52	12.68	0.95	5.47	66.58	1.4
C-9	47.33	5.71	13.08	1.15	5.00	33.56	1.4
Lake Amadeus-1							
LA-1–5	65.55	5.87	3.57	0.76	12.0	21.93	–C

The abundance of pyrite is evident in reflected light (Figs. 3.2 and 3.4). Only traces of Fe were detected by microprobe analyses within humic acids seen in TEM. Most opaque particles within humates analysed by EDS gave Cu peaks.

Yet another indication that the majority of the iron is not concentrated in the organic fraction.

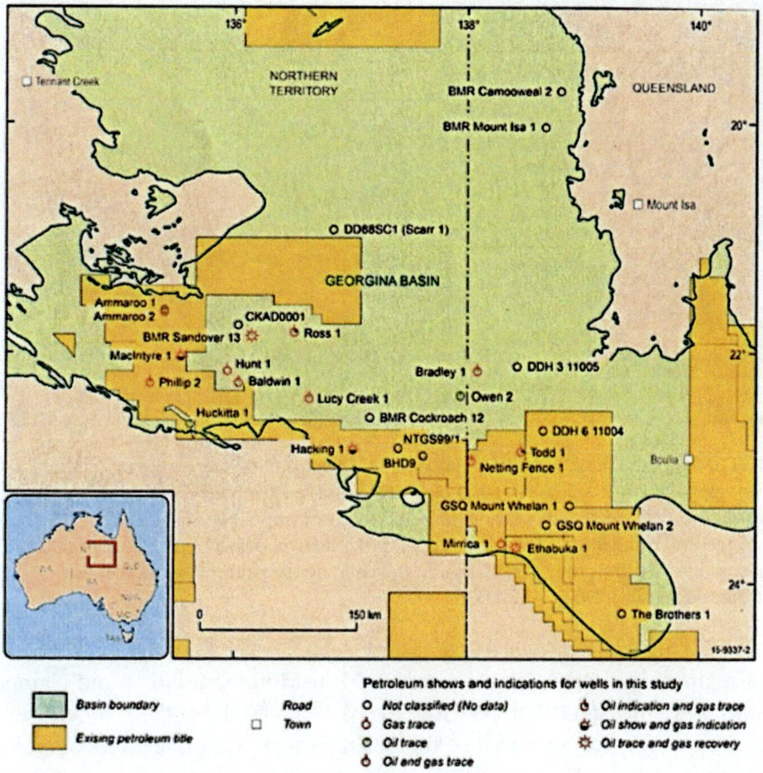

Map showing borehole locations in Georgina Basin. *Source* Geoscience Australia

The reflectance (Ro) profiles of drill holes from the Middle Cambrian of the southern part of Georgina Basin display a typical profile of OM that has been subjected to hydrothermal heat effect (Glikson and Golding 1998; Glikson 2001; Glikson et al. 2006). Reflectance measurements were taken from the predominant OM in each sample. The perturbations of the reflectance profile occurred at a certain level, whereas above this level the maturation profile was 'normal', namely gradual increase with depth when the heat flow was from the basement in a constant geothermal regime. Most boreholes were below the level deemed 'normal' as evident from the Ro data of algal remains. Only few boreholes started with low Ro OM (0.2–0.4%) over a very limited depth. Borehole Scarr which is in the northernmost part of the southern basin and bordering on the northern thermal regime of the basin shows a 'normal' maturation profile with depth. Although gradual increase in Ro follows to 320 m; A sharp increase of Ro from 0.4 to 0.8% (from >40 to 140 °C) occurs abruptly over a mere 10 m depth, indicating entrance into a significantly higher

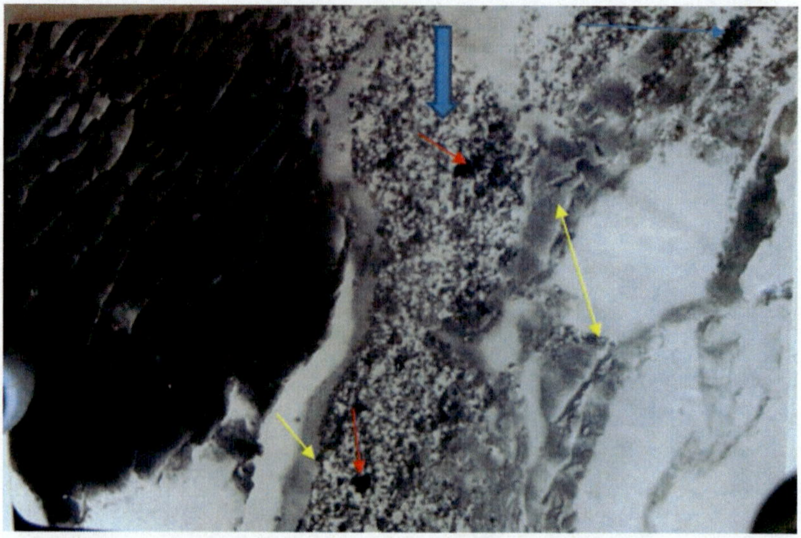

Fig. 3.6 TEM showing high reflectance (Ro%) OM, electron opaque, on the far left, next to humic acids (tiny globules, blue arrow) wedged between dense, electron opaque OM and cyanobacterial filaments of low electron density OM on the right (yellow arrow). *Note* Electron density showing in reverse to picture in reflected light microscopy; dark electron dense = high Ro, bright in reflected light microscopy; electron transparent, lipidic (L) band on the right = low Ro. Humic acids absorb metals (thin red arrows); Mag. ×130,000

thermal regime. In contrast predominant OM in Mount Isa No. 1 and Camooweal Boreholes was consistently of very low Ro (0.1–0.4%). Both these boreholes are in the far north of the northern part of the basin. Noteworthy is Brothers borehole, at the southern end of the Georgina Basin with OM displaying strong perturbations and reaching highest Ro (highest temperature).

The hydrothermal heat effect in the southern part of the basin is noted by a sudden jump in the Ro and an erratic Ro profile below that point (Fig. 3.7a–e). Furthermore, the OM in southern and south-western Georgina Basin borehole samples contained solid hydrocarbons, both bitumen and pyrobitumen, the residue of several generations of oil. Low Ro OM in TEM showed algal structures (Glikson et al. 1992). The same study showed the high Ro OM in TEM having spongy structures characteristic of algal remains after oil generation. There is evidence of several hydrothermal events that brought on generation of oil, the light fraction would have moved away, into reservoirs while the heavy oil solidified into bitumen. The bitumen became pyrobitumen following another later hydrothermal event. Similar sequence of events had been reported from mount Isa and Mc Arthur basins (Glikson et al. 2006). The southern part of Georgina Basin by being close to major fault systems was in a regime of hydrothermal surges at times which generated significant quantities of oil and gas

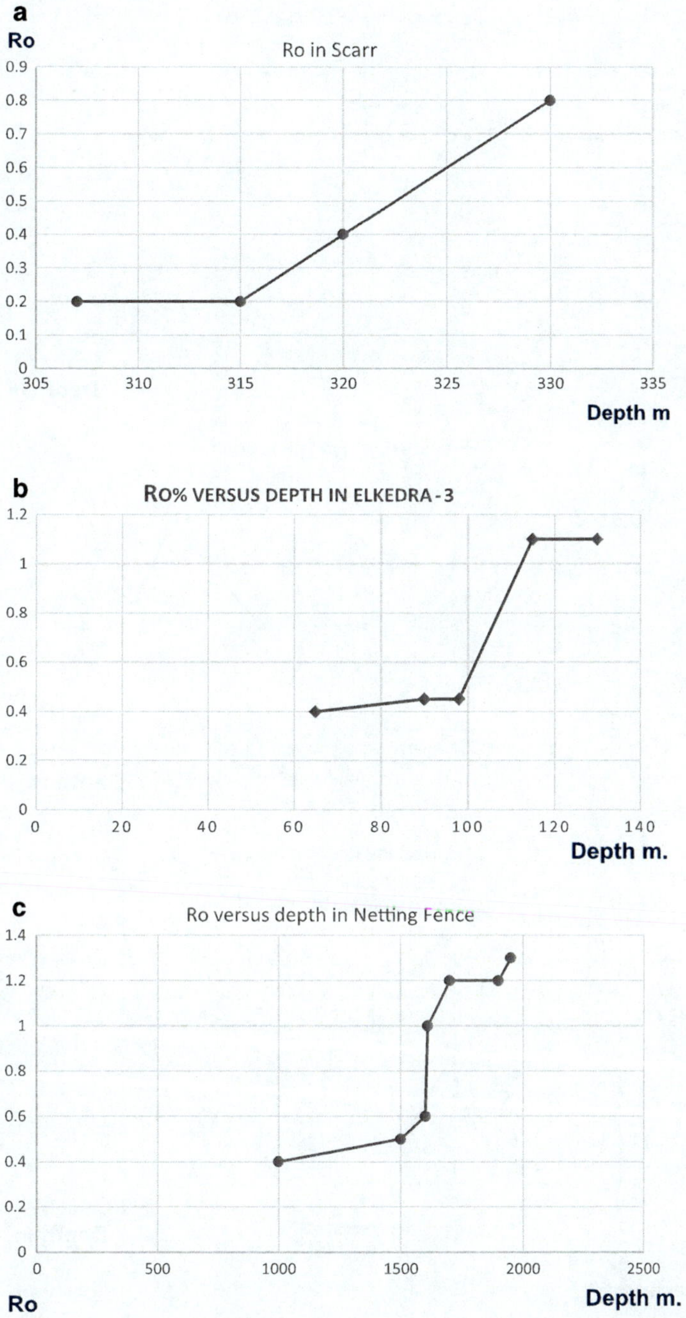

Fig. 3.7 a–e Reflectance profiles of boreholes in Middle Cambrian of south and south west Georgina Basin. Ro maximum temperature calculated using Aizawa's (1990) conversion scale

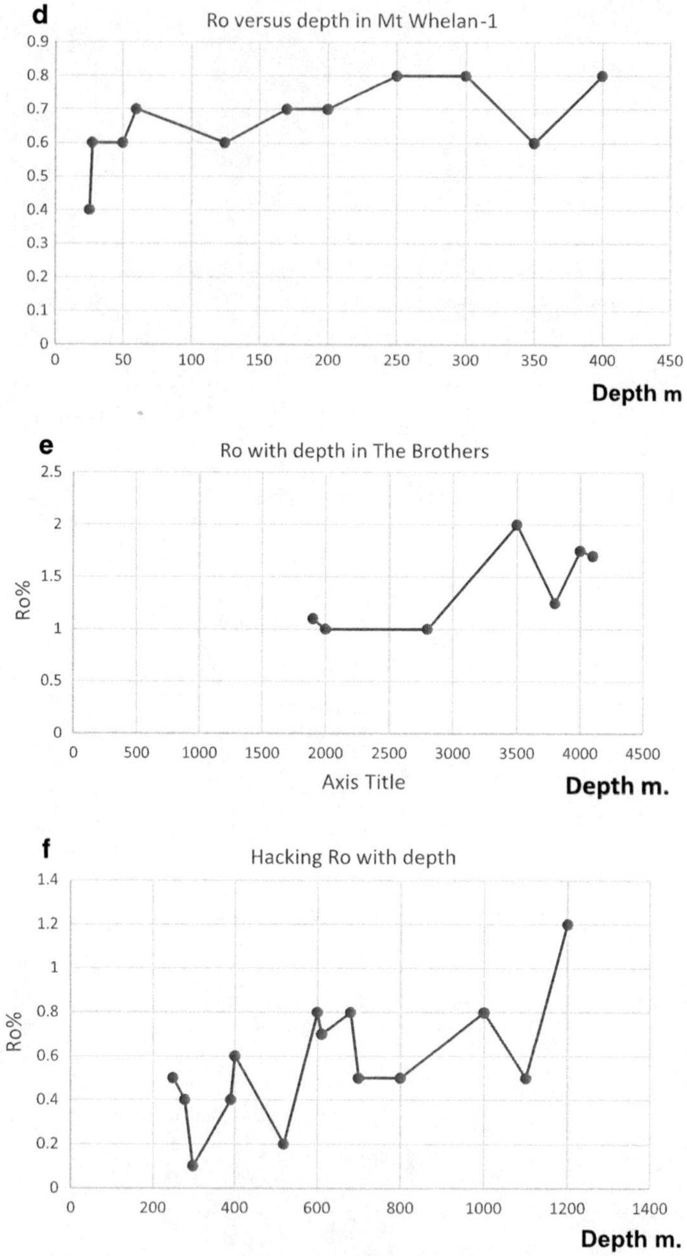

Fig. 3.7 (continued)

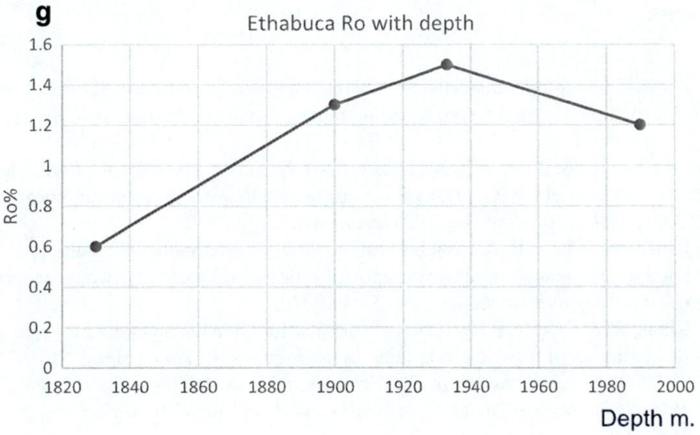

Fig. 3.7 (continued)

from the rich oil shales. These hydrothermal effects on the OM are not evident in the northern part of the basin where the main OM is of low Ro and bitumen is absent. Therefore, the deposits in the northern part of Georgina Basin are oil shales with a potential to generate oil. These oil shales yielded 100 l/ton on heating artificially (Glikson et al. 1985).

Greene (2010) described the faulting and rifting processes in southern Georgina Basin, as two stage occurrences: major rifting occurring in approximately 700–650 Ma and a reactivation of the faults at around 600 Ma. These faults were conduits for hot fluids that heated the OM in the southern parts of the basin and generated oil and gas at times. Numerous minor fault systems are associated with some of the intrusions in the south western and eastern parts of the southern Georgina basin (BMR Geological Maps, in Smith 1972).

Boreham et al. (2014) identified three depositional environments in the Cambrian of southern Georgina Basins:

1. Early Middle Cambrian anoxic platform.
2. Late Middle Cambrian euxinic, deep-water facies.
3. Early–Late Cambrian anoxic evaporitic sabkha facies.

Boreham's distinct facies help to understand the observed diverse OM types in the Cambrian sediments; cyanobacterial remains are the dominant facies in the northern Georgina Basin, while in the southern parts of the basin algal remains prevail.

Khan et al. (2019) described organic matter compositions representing two facies, from the Lower Cambrian of the Salt Range in Pakistan closely resembling those in Georgina Basin; algae (planktonic) and cyanobacterial mats deposited under 'stratified conditions in a shallow restricted basin dominated by cyanobacteria and sulphate reducing bacteria'.

References

Aizawa, J. (1990). Palaeotemperatures using fluid inclusions and coal rank of Carbonaceous material of Tertiary formations in south-east Kyushu, Japan. *Japanese Journal of Earth Sciences, 85,* 145–154.

Boreham, C. J., Newman, J., & Edwards, D. (2014). Maturity assessment of Middle Cambrian organic-rich rocks in Southern Georgina Basin using geochemical and petrographic comparisons. In *AOGC 2014, 18th Australian Organic Geochemistry.*

Conrad, R., Klose, M., Claus, P., & Enrich-Prost, A. (2010a). Methanogenic pathways, ^{13}C isotope fractionation and archaeal community composition in the sediments of two clear water lakes in Amazonia. *Limnology and Oceanography, 55*, 689–702.

Conrad, R., Claus, P., & Casper, P. (2010b). Isotope fractionation during methanogenic degradation of OM in the sediment of an acidic bog lake. *Limnology and Oceanography, 55*, 1932–1942.

Cook, A. C. (1982). Organic facies in the Eromanga Basin. In *Eromanga Basin Symposium: Summary Papers, Geological Society of Australia and Petroleum Exploration Society of Australia* (pp. 234–255).

Creveling, J. R., Johnston, D. T., Poulton, S. W., Kotrc, B., März, C., Schrag, D. P., & Knoll, A.H. (2014). Phosphorus sources for phosphatic Cambrian carbonates. *GSA Bulletin, 126*(1/2), 145–163.

Glikson, M. (2001). The application of electron microscopy and microanalysis in conjunction with organic petrology to further the understanding of organic-mineral association: Examples from Mt Isa and McArthur Basins, Australia. *International Journal of Coal Geology,* 139–159.

Glikson, M., & Golding, S. D. (1998). Organic matter maturation as an indicator of hydrothermal processes in sedimentary basins. In G. B. Arehart & J. R. Hulston (Eds.), *Water rock interaction* (pp. 101–104). Rotterdam: Balkema.

Glikson, M., Gibson, D. L., & Philp, R. P. (1985). Organic matter in Australian Cambrian oil shales and other Lower Palaeozoic shales. *Chemical Geology, 51,* 175–191.

Glikson, M., Taylor, D., & Morris, D. (1992). Lower Palaeozoic and Precambrian petroleum source rock and the coalification path of alginate. *Organic Geochemistry, 18*(6), 881–897.

Glikson, M., Golding, S. D., & Southgate, P. N. (2006). Thermal evolution of the ore-hosting Isa Superbasin: Central and Northern Lawn hill platform. *Economic Geology, 101,* 1211–1229.

Golubic, S., & Abed, R. M. M. (2010). Entophysalis mats as environmental regulators. In J. Seckbach & A. Oren (Eds.), *Microbial mats: Modern & ancient microorganism in stratified systems* (pp. 232–251). Springer.

Greene, D. C. (2010). Neoproterozoic rifting in the southern Georgina Basin, central Australia: Implications for restructuring Australia in Rodinia. *Tectonics, 29*(5), 46 pp.

Hayes, J. M. (1993). Factors controlling ^{13}C contents of sedimentary organic compounds: Principles and evidence. *Marine Geology, 113,* 111–125.

Khan, I., Zhong, N., Luo, Q., & Ai, J. (2019). Neoproterozoic: Lower Cambrian organic-rich shales of Salt Range Formation, Upper Indus Basin, Pakistan. *International Journal of Coal Geology, 217.*

McKirdy, D. M. (1977). *Diagenesis of microbial organic matter* (PhD thesis). Dept. of Geology, Australian National University, Canberra, Australia.

Oehler, D. Z. (1978). Microflora of the Middle Proterozoic Balbirini dolomite (McArthur Group of Australia). *Alcheringa,* 269–309.

Riding, R. (1991). Cambrian calcareous cyanobacteria and algae. In R. Riding (Ed.), *Calcareous algae and stromatolites* (pp. 305–334). Springer.

Sandstrom, M. W. (1982). *Organic geochemistry of phosphorites and associated sediments* (PhD thesis). Research School of Earth Sciences, The Australian National University, Canberra, Australia.
Smith, K. G. (1972). Stratigraphy of the Georgina Basin. *Bureau of Mineral Resources, Geology and Geophysics Bulletin, 111,* 1–156.

Chapter 4
Julia Creek Oil Shales

Abstract The term Julia Ck. has been given to shales, siltstones and mudstones with organic carbon occasionally reaching over 50%. More often it is less. The organic-rich horizons within the Lower Cretaceous Toolebuc Formation are in the northern part of the Eromanga Basin (Fig. 4.2) which comprise the oil shale facies. The oil shale facies is characterised by organic-rich beds alternating with calcite. The calcareous content varies from 10% in boreholes Boulia-3, Longreach-1 and Tambo-38 to more than 50% in Jericho-11 borehole. The organic matter is lamellar intercalated with calcite. Shells and calcareous remains of foraminifera, coccoliths and cyanobacteria are common within the oil shale beds. The northern part of the Eromanga Basin where the oil shale facies is recognised was a partly enclosed sea in the Lower Cretaceous with only its top part of the water column having access to the open Carpentaria Basin in the north, thus having restricted water circulation to upper water levels only, creating the anoxic conditions at the water–sediment interface and immediately above it for the preservation of OM. These deposits have been extensively evaluated and considered for mining as oil producing oil shales. The following chapter will demonstrate their 'poor quality' in terms of oil shale resource.

4.1 The Julia Creek Oil Shales

The Julia Creek oil shales are a part of the Toolebuc Formation in the Eromanga Basin (Senior et al. 1975; Exon and Senior 1976; Ozimic 1983). The Formation is of Lower Cretaceous age and contains oil shales in the northern parts of the Eromanga Basin (Fig. 4.2). Restriction of oil shale facies to the northern part of the Eromanga Basin is related to the exclusive anoxic conditions in that part of the basin. The difference is marked by the absence of benthic foraminifera in the southern basin, lack of anoxic bottom waters and presence of planktonic organisms dominated by dinofalgellates. Oil shale facies sediments in the Toolebuc Fm. are characterised by OM alternating with calcite. The amount of calcareous matter varies from about 10% to over 50%. The OM is compressed into lamellae and intercalated with calcite (Fig. 4.1). Complete shells of *Inoceramus* and *Aucellina* are common in some horizons. Away from the oil shale facies, as observed in boreholes Augathella and Tambo the predominant

M. Glikson-Simpson, *Oil Shales*, https://doi.org/10.1007/978-3-030-60675-6_4

rock is a black or dark grey silty shale with occasional land plant remains and few *Aucellina* shells.

In the present study, polished whole rock samples showed the OM to be vitrinite-like, of lamellar appearance (Fig. 4.3), low Ro (0.2%) and intercalated with calcite. In TEM it appears as degraded OM devoid of defined structure (Fig. 4.12). The organic laminae interbedded with calcite are suggestive of cyanobacterial mats. However even TEM observation did not show definite identifiable structures in the OM except bundles of microbial filaments (Figs. 4.10; 4.11) and other bacterial forms (Figs. 4.13; 4.14; 4.15). The intense degradation of the primary OM obliterated any visible structures (Figs. 4.12).

Pyrite abounds in massive and framboidal form within OM, within calcareous remains of organisms and within the crystalline calcite. A wealth of planktonic foraminifera and coccoliths are visible in polished whole rock surfaces observed in reflected light (Figs. 4.3 and 4.5b) It appears from these observations that the planktonic organisms fell into and were deposited in soft Cyanobacterial mats, displacing and warping these mats as can be seen from their discordant positions in respect to the lamellar structure (Fig. 4.3).

The main mineral component is calcite which is either closely associated with OM or is in the form of remains of calcareous organisms such as foraminifera, coccoliths (Figs. 4.6; 4.9) and cyanobacteria (Figs. 4.7; 4.8). A characteristic feature is the calcite—organic matter close association, expressed in the irregularity in thickness and extent of the organic laminae entwined with calcite.

When using ultra-violet light with blue filter excitation, the only fluorescence response comes from occasional spores and pollen of land plants and dinoflagellates. The predominant vitrinite-like OM lacks any fluorescing properties, and when observed in transmitted light is amorphous, devoid of any definite structure that can be identified with an organism or a plant.

Fig. 4.1 Julia Creek oil shales; black laminae are organic matter. *Note* OM: Mineral ratio in favour of minerals, mainly calcite

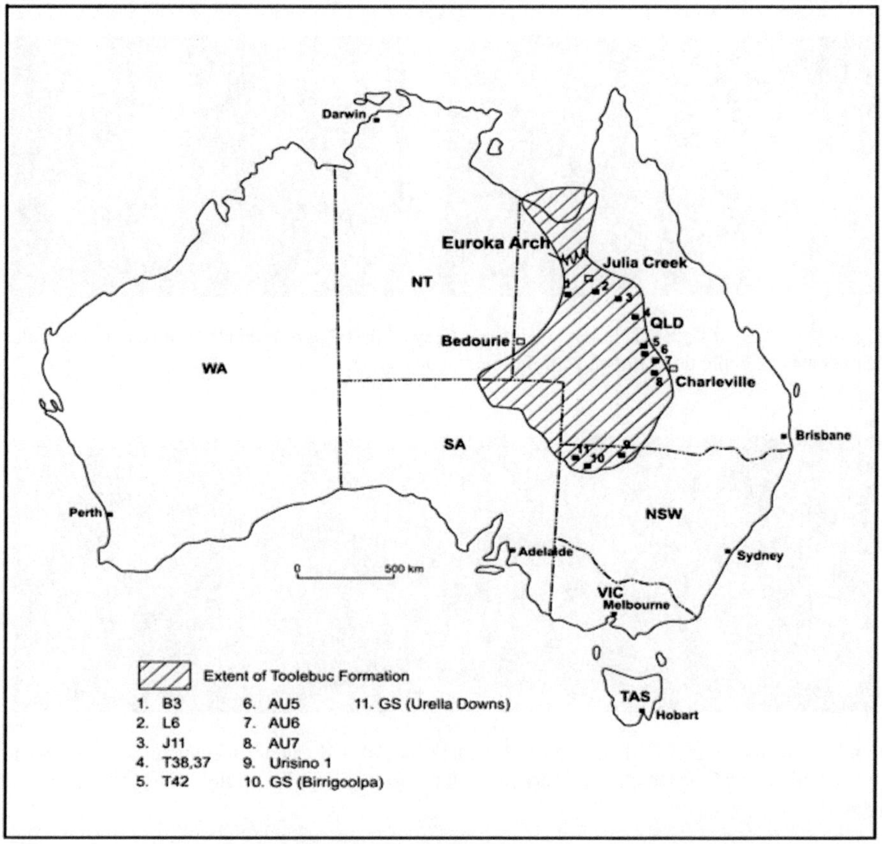

Fig. 4.2 Julia Ck oil shales, Toolebuc Formation and location of Boreholes. Source: Ozimic 1982

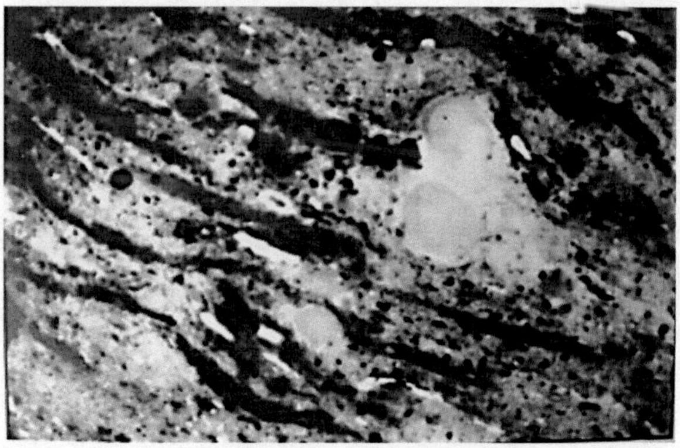

Fig. 4.3 A mould of Calcareous remains of a foraminifer displacing banded OM. Dark particles are pyrite. Reflected light microscopy. X100.Note distortion of laminae as calcareous organisms fell into OM

a b

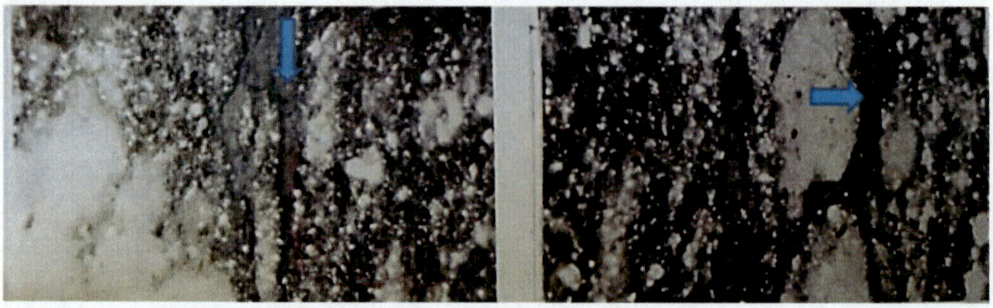

Fig. 4.4 **a–b**; Bands of OM (Black low Ro and Grey, higher Ro) intercalated with calcite. Spicules of tunicates as while dots. Reflected light microscopy. X100

a b

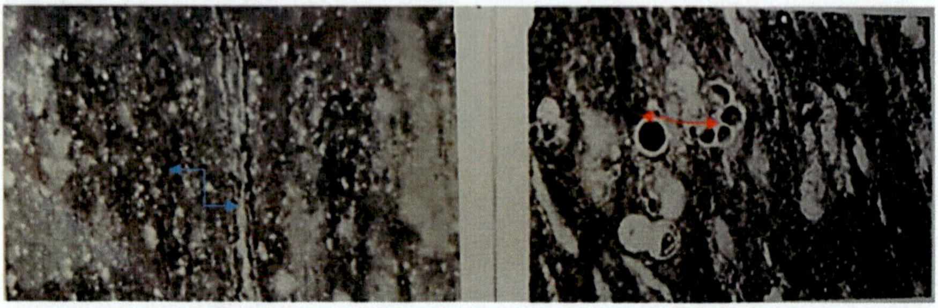

Fig. 4.5 **a** Fine bands and lamellae of OM (dark bands) and calcite; **b** remains of foraminifera mixed with OM. **c** Fine lamella of higher Ro OM. **d** Foraminifera shells filled with pyrite. Reflected light. X 100

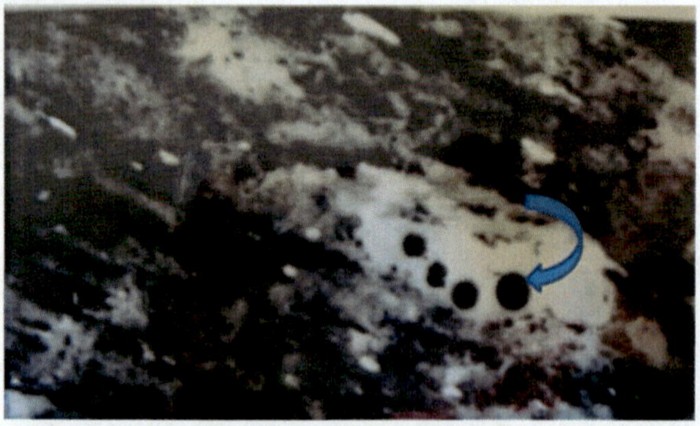

Fig. 4.6 Calcareous concretion with pyrite. Reflected light. X100

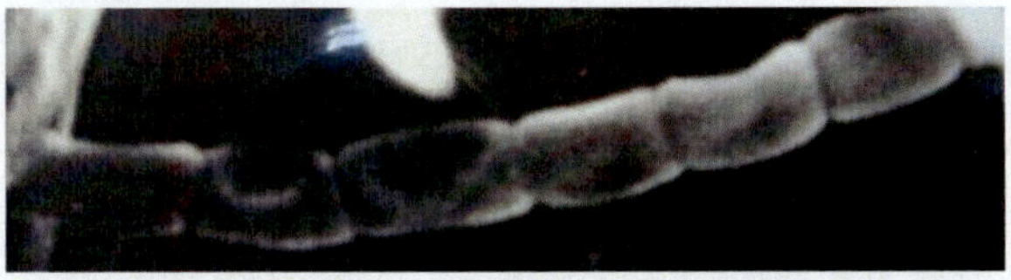

Fig. 4.7 SEM of cyanobacterial skeletal filament (courtesy Scheibnerova 1982) X 15 000

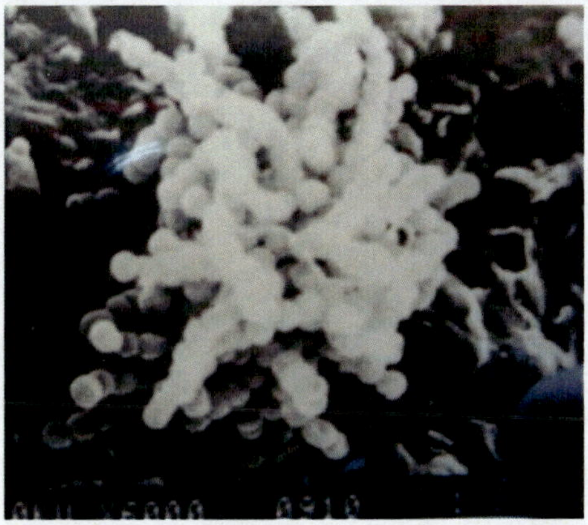

Fig. 4.8 SEM of skeletal cyanobacterial colony (Photo courtesy, Scheibnerova 1982) X 6 000

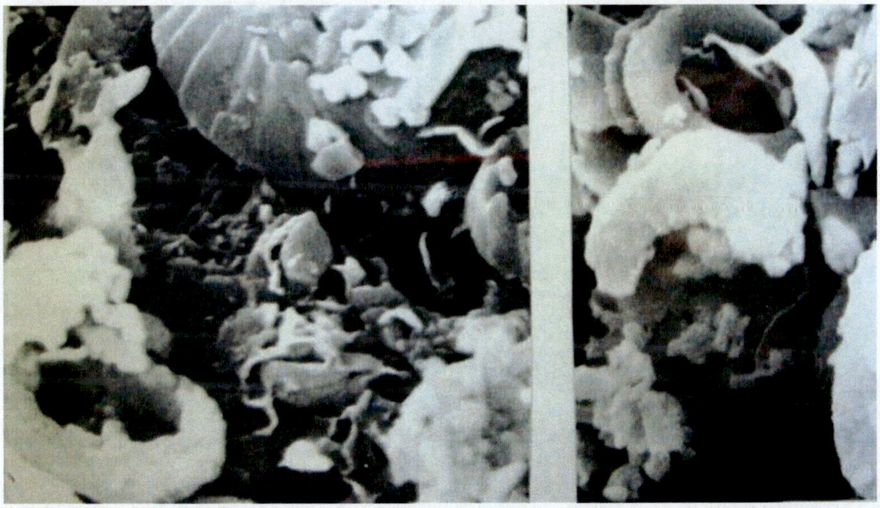

Fig. 4.9 SEM of Coccoliths in Toolebuc Fm. X8 800

Fig. 4.10 TEM of filamentous microbial remains, possibly the organic remains of cyanobacteria. Mag.X80 000

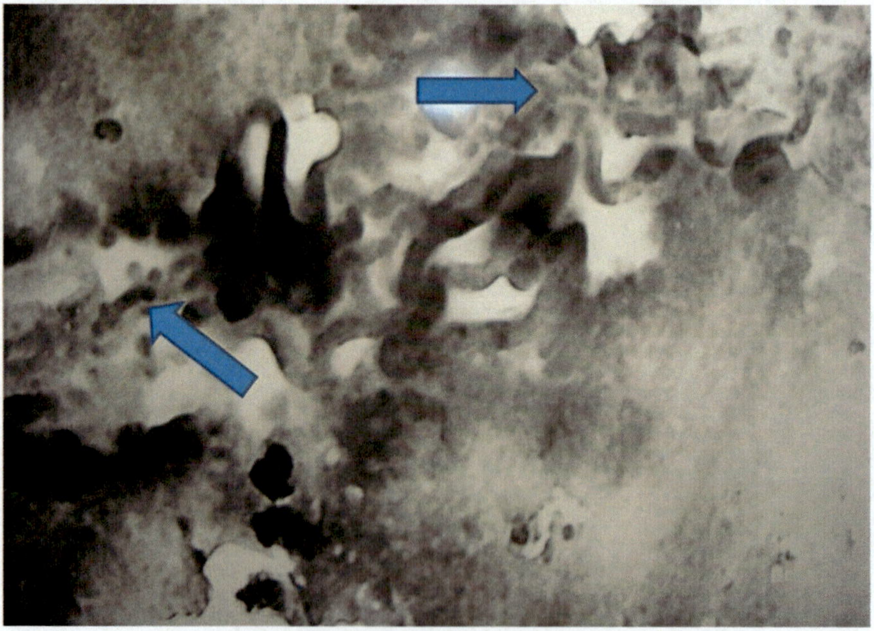

Fig. 4.11 TEM of filamentous bacteria. The bacterial bundles are seen disintegrating into particles half-way to becoming humic acids (thick arrow) X132 000

4.1.1 TEM Observations

TEM observations of ultra-thin sections of OM revealed for the most part, OM of relatively low electron density (light grey in appearance) and somewhat irregular surface, almost sponge-like, unlike OM sourced from green algae which has a smooth homogenous surface as seen in Torbanites and Coorongite, and in the Green River oil shales. The characteristic micro-globules of humic acids are seen mostly scattered irregularly throughout the rest of the OM (Fig. 4.12). Highly electron dense particles occasionally associated with the humic acids when analysed by EDS were iron. Cu was detected in all the electron dense coccoid bacterial forms. Filamentous organisms, possibly the soft parts of cyanobacteria can be recognised in many samples (Figs. 4.10; 4.11). Single filaments are rarely distinguishable as the filaments mostly merge into loosely compacted bundles. Most of the filaments do no longer show any internal structure; only on a rare occasion cross walls could be recognised in TEM. The cross walls are clear in SEM of the calcareous skeletons of the cyanobacteria (Fig. 4.7; 4.8). The filaments are non- branching and straight or curved. Individual filaments are 0.25μ wide and 1–2 μ long.

SEM observations of whole rock fragments showed calcareous remains of organisms closely resembling the cyanobacterium *Oscillatoria* (Fig. 4.7). The main difference between the cyanobacterial remains observed in the Toolebuc Formation oil shales and modern cyanobacteria is in their size. Modern cyanobacteria are visible in optical microscopy, whereas the Cretaceous forms are visible only under high magnification of electron microscopy. The Difference in size between the Cretaceous fossil forms and their modern counterparts has been discussed in an earlier paper (Glikson and Taylor 1985). The studies of Malone et al. (1975) and Pardue et al. (1976) have been mentioned. Their studies demonstrated that gradual depletion in Nitrogen with population 'explosion' of cyanobacteria leads to decreasing size of individuals within the population.

TEM observations of the organic components in the Toolebuc oil shale showed the predominant OM to be of low electron density, porous where occasional filamentous structures could be recognised. The humic acid fraction is small, rarely more than 1–2%. Recognisable microbial remains were observed in numerous samples. One sample showed clumps of a stalked microorganism resembling *Caulobacter*. This organism has been known to be associated with cyanobacteria and forms rosettes that aggregate into clumps up to 2μ across (Figs. 4.14 and 4.15). It has been shown experimentally that *Caulobacter* microorganisms tend to aggregate and clump into rosettes when stressed (Martinez et al. 2006). The *Caulobacter*-like microbes appear in association with highly electron dense coccoid forms (Fig. 4.14) resembling microbial remains such as have been reported by Degens and Ittekot (1982) from the OM in the Cretaceous Messel oil shales. The electron density of these forms was attributed to metal bonding by the bacterium's cell walls. EDS analysis of the coccoid forms in the Toolebuc oil shales displayed mainly Cu peaks, and only minor peaks of Fe and Mg. A microorganism closely resembling *Sulfolobus* has been observed in some samples (Fig. 4.13). Individual forms of this bacterium are spherical and doughnut shaped when sectioned, with one convex side, closely resembling the living deep

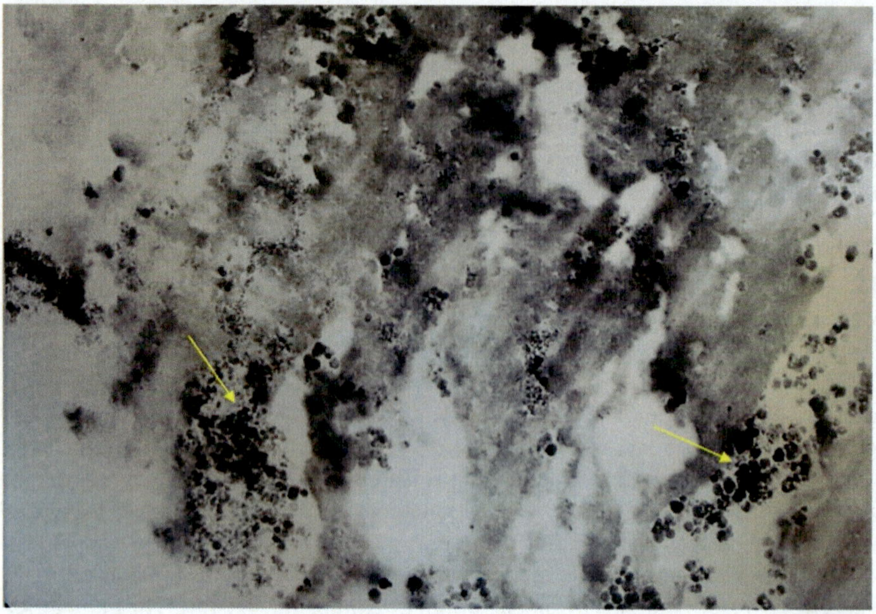

Fig. 4.12 TEM shows predominant OM (vitrinite of low Ro in reflected light) making up the bulk of the background, suggestive of degraded microrganism remains. Humic acids (arrows) distributed throughout the matrix OM, often electron opaque due to absorbed metals. X33 000

sea bacterium *Sulfolobus,* a sulphate-reducing microbe, known to be associated with Cyanobacteria.

The predominant very low Ro vitrinite does not fluoresce, and when viewed in transmitted light microscope is amorphous, namely devoid of any distinguishable structures, characteristic of degraded OM.

Calcareous outlines or replicas of *Oscillatoria* –like cyanobacteria and other calcareous microorganisms were observed in SEM of whole rock samples (Figs. 4.7 and 4.8). The rarity of this kind of preservation is explained by Berner (1968) in terms of very high rates of calcite precipitation; where Ca salts of fatty acids are formed prior to calcite and the transformation from the one to the other involves a considerable increase in density causing the destruction of the external morphology, so what is left is just a mould. The irregularity of the calcareous lenses and laminations expressed in variability of thickness and lateral extension cannot be explained solely as a result of seasonal changes in the cyanobacterial communities. They are more likely the result of localised micro-environmental changes in the cyanobacterial mats. This observation is supported by the isotopic compositions of the OM-associated calcite compared to the isotopic composition of the benthic shells.

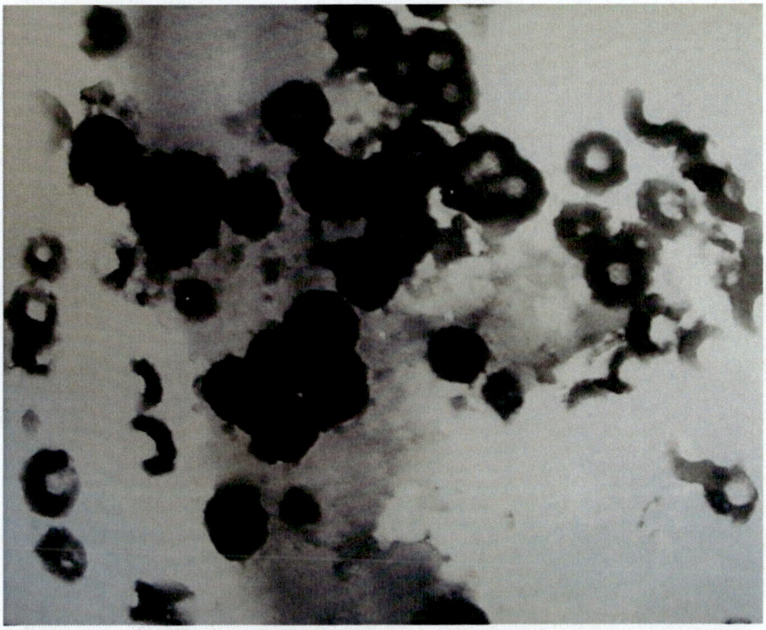

Fig. 4.13 TEM showing bacteria in Toolebuc oil shale closely resembling *Sulfolobus* embedded in the main OM. (Mag.X 33 000)

4.1.2 C-Isotope Composition of Julia Ck. Formation Oil Shales

It is evident from the carbon isotope results that values of shells analysed differ significantly from values obtained from the calcareous matter intercalated with OM. These results point to two different carbon sources: (1) Carbon from CO_2 released by the decomposing cyanobacteria, already depleted in ^{13}C providing carbon for $CaCO_3$ precipitating in the mat; and (2) a carbon source being primarily the bicarbonate of sea water from which the *Inoceramus* shells derived their carbon. The latter is therefore isotopically 'heavier' than the carbon incorporated in the Calcite precipitate (Table 4.1).

Irwin et al. (1977) found evidence for diagenetic carbonates from organic-rich sediments, and concluded that CO_2 released from decomposing OM dissolves readily in pore water to raise the pH and bring about $CaCO_3$ precipitation. These authors also expected 'lighter' C-isotope values in these carbonates. The values obtained by Irwin et al. showed a more pronounced difference between the two carbonates than the results obtained from the Toolebuc samples. The reason for the smaller difference between the shells and diagenetic calcite in the present study is thought to be in the technical difficulty of separating the $CaCO_3$ precipitate intimately associated with OM, from the coccoliths shells and the planktonic foraminifera. However, there is

Fig. 4.14 *Caulobacter*—like microbial 'rosettes' associated with coccoid methanogens Mag.X10 400

enough evidence to suggest two different carbon sources: The OM-associated $CaCO_3$ derived its CO_2 from decomposing cyanobacteria already depleted in ^{13}C, whereas *Inoceramus* has derived its carbon from the bi-carbonate of the sea-water reservoir.

Carbon isotope composition of $CaCO_3$ of *Inoceramus* is within the range obtained by Degens and Epstein (1964) for hypersaline environments. The Toolebuc sea was partly divided from the open sea in the north by the St Elmo rise which would have induced higher salinities in bottom waters of a stratified sea.

The C-isotope composition of OM (Table 4.2) from oil shale bearing horizons ranges between −27 and −29‰. In interpreting these results, the following factors are considered: Low ^{13}C values may be the result of concentration of lipidic matter which is resistant to degradation and tends to be preserved in sediments (Eadie et al 1978; Teichmuller 1974; Glikson-Simpson 2020). Another factor influencing the significantly high ^{13}C depletion in the Toolebus oil shale samples may be the amount of CO_2 available in the immediate environment of the cyanobacterial mats. It was demonstrated by Pardue et al (1976) that an increase in CO_2 supply to cultures of cyanobacteria was followed by increase in fractionation coinciding with

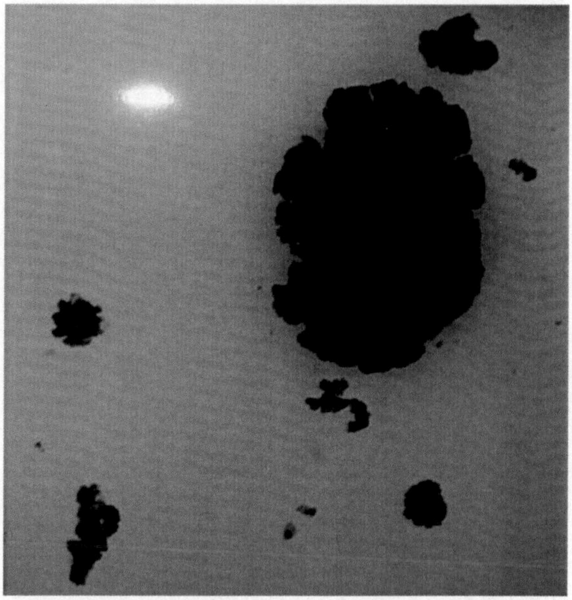

Fig. 4.15 A large Rosette of *Caulobacter*-like microbes. X104 000

Table 4.1 C-isotope composition of diagenetic calcite and *Oucellina/Inoceramus* shells in comparison

Sample no	$\delta^{13}C$‰ of shells	$\delta^{13}C$‰ of calcite
B3-5	+ 0.20	−0.60
B3-6	+ 1.25	−0.21
"	+ 1.80	−0.20
L6-8	+ 0.10	−0.60
L6-1	+ 1.17	−0.88
B3-6	+ 0.39	
B3-7	+ 0.27	
L6-1	+ 0.84	−1.66
L6-2	+ 0.28	
L6-3	+ 0.26	
J6-8	+ 0.10	

maximum metabolism. T. Sharkey (pers.comm.) obtained values of $\delta^{13}C$ around −30‰ and lower for cyanobacteria cultured in CO_2-enriched atmosphere. Francey (1982) reported values of $\delta^{13}C = -24$‰ in trees where CO_2 was collected from the canopy of the forest, and −29‰ in trees under the canopy, where CO_2 levels were significantly higher than in the canopy.

It is evident that high CO_2 concentration in the cyanobacterial mats is a factor in obtaining the ^{13}C-isotope depleted values in the Toolebuc oil shales. Recent cyanobacterial mats are considerably isotopically 'heavier' than Toolebuc values.

Table 4.2 Isotope composition of OM in Toolebuc Fm. Oil shales

Well and Sample NO	Depth in m	Fraction	$\delta^{13}C$‰
Tambo 38/10–2	54.0	OM	−27.75
“ “/11–1	56.0	OM	−26.35
“ “/11–2	57.0	OM	−27.61
“ “/ 1–3	58.0–59.0	OM	−27.60
“ “/11–3	OM less HA	−28.40	
“ /11–3	HA	−28.40	
“ / 6–3	OM	−28.86	
“ / 6–3	HA	−25.60	
“ / 2–3	61.0	OM	−23.00
“ / 9–3	63.0	OM	−27.99
Boulia 3/B3-2	30.0	OM	−27.27
“ “/B3-3	30.5	OM	−28.40
“ “/B3-5	31.0	OM	−28.14
“ “/B3-1	31.4	OM	−29.00
“ “/B3-4	32.0	OM	−27.80
“ “/B-6	32.3	OM	−28.15
Jericho11/J11-11	54.0	OM	−27.12
Longreach 6/L6-2	25.3	OM	−27.90
“ “/L6-12	28.0	OM	−27.19
Augathella 6/AU6-40	32.0	OM less HA	−25.10
“ “/Au6-40	32.0	HA	−22.90
Julia Ck./A	OM less HA	−25.35	
“ “ /B	HA	−25.56	
Urella Downs/GS-6	32.0	OM	−22.00
“ “ /GS-4	120.0	OM	−23.60
Petita Bore/GS-5	301.0	OM	−23.31
Birrigoolpa/GS-8	127.0	OM	−25.70

OM = organic matter
HA = humic acid fraction (highlighted)

Calder and Parker (1973) reported $\delta^{13}C$ values of −13 to −17‰ for cultures of cyanobacteria. Behrens and Frishman (1971) studied carbon isotope compositions of recent cyanobacterial mats from a lagoon in Baffin Bay, Texas, and obtained values of −14 to −16.7‰. However, with onset of degradation the depletion of the OM in respect to ^{13}C is expected to increase.

The dark, anaerobic metabolism of certain mat-forming cyanobacteria, such as *Oscillatoria* may bring about further depletion in ^{13}C, by as much as 5‰ (Degens 1969). The activity and input from methanogenic and sulphate reducing bacteria is another significant factor in obtaining isotopically ‘lighter’ values. They appear to be heavily involved in the breakdown of OM in these oil shales. The common sulphate reducing bacterium *Desulfovibrio desulfuricans* known to be always associated with cyanobacteria was noted to preferentially metabolise ^{12}C.

Some samples from Drill holes Tambo, Boulia, Jericho and others had negative isotope values of OM similar to what is usually obtained from coals or other terrigenous sources. Input from terrigenous source at times was brought into the marine environment by rivers flowing into the bay during a time of heavier rains.

Samples where the humic acid fraction has been analysed separately from the total OM, the C isotope compositions were the same or only slightly 'heavier' than the bulk OM in the same samples. These results indicate the humic acids being predominantly insitu formed from amino acids and carbohydrates released by the decomposing cyanobacterial mats, terrigenous OM input and degradation products of other organisms. The three samples from Urella Downs and Petita yielded isotopically heavier values than all other bore holes. These samples came from drill holes on the southernmost margins of the basin where no oil shales are present and the OM yielded isotopic values such as have been reported to be characteristic of plankton (Nishimura 1977).

Salinities assumed for the upper part of the water column or water surface of the Toolebuc sea were in the vicinity of 35‰, which would provide optimum conditions for the coccoliths (Fig. 4.9) which are extremely abundant in this formation, pointing to optimum conditions for their growth shown to be salinities exceeding 34‰ (Hubert and Rodan 1963). Optimum temperatures for the growth of the common marine coccoliths, *Coccolithus huxleyii* is in the 18–24 °C, and pH around 8. The vast amount of calcified coccoliths in the Toolebuc oil shales points to optimum conditions prevailing. Bottom waters were denser and more saline as would have suited the cyanobacterial mats which thrived at the water—sediment interface. High salinities are not only tolerated by cyanobacteria, but they flourish in environments with salinities ranging from 45–50‰ (Spencer Gulf, South Australia; Shark Bay, Western Australia) to hypersaline (Solar Lake, Red Sea). As to oxygen in the lower parts of the water column, it has been shown (Ryan and Cita 1977) that often only ¼ of the oxygen percolates to the bottom of the sea. No direct evidence is available in regard to temperatures, only indirect assumptions can be made based on the existence of cyanobacterial mats in the environments mentioned above.

The abundance of pyrite closely associated with the organic matter and within calcareous remains of micro-organisms points to anoxic conditions at the sediment—water interface, conditions which ultimately enabled the preservation of the OM. Sulphate reducing bacteria are known to be intimately associated with cyanobacteria (Bauld et al. 1980, 1981). Various bacterial remains are visible in TEM, some of which are suggestive of sulphate reducing microbial remains.

4.1.3 Major Elements

Organic carbon content in Toolebuc Formation oil shales ranges between 8 and 66%, with the majority of samples below 30% which is not favourably comparable to the 'real' oil yielding sediments such as Torbanites, Kukersite, Tasmanite and Green River, Rundle Complex or even the Camooweal oil shales. Usually carbon and hydrogen make up the bulk of the OM (Tissot and Welte 1978) whereas oxygen is the next most abundant element, particularly in the humic acid fraction of the OM. Nitrogen and sulphur vary. Inorganic sulphur occurs in pyrite and can be seen in light microscopy and clearly in TEM, where it is observed as electron opaque spots in the humic acid fraction. This is expressed in the graph of Fig. 4.16 as the high S is predominantly positively correlated with organic carbon. On the other hand, observations in light microscopy show pyrite particles, from minute to larger concentrations present in the organic as well as calcareous fractions of the oil shales. Clearly there are two sources of sulphur; sulphides in pyrite and organic sulphur tied up in the humic acid fraction, the degraded product. High organic sulphur is typical of cyanobacteria- dominated OM where sulphate reducing bacteria are active.

Nitrogen is usually obtained from humic-acid rich samples associated with cyanobacteria. Carbon is progressively concentrated with increasing degree of maturation.

The classical Van-Krevelen diagram developed for coals is adopted for all other OM including oil shales. By expressing the ratio between the elements carbon, hydrogen and oxygen, three types of 'kerogen' (OM) have been designated (Tissot and Welte 1978), based on the concentration of the three elements relative to each other: Type I OM has a high hydrogen content (H/C = 1.5 or more) as a result of

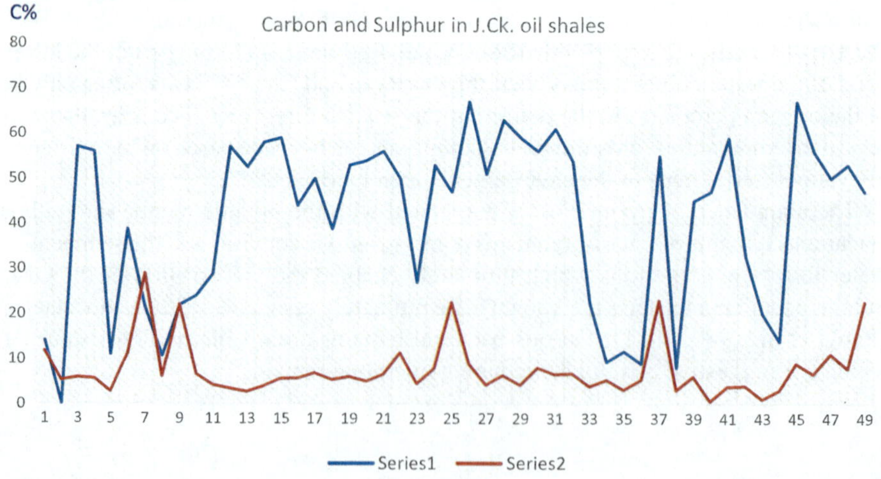

Fig. 4.16 Relationship between Carbon (series 1) and Sulphur (series 2) in Toolebuc/Julia Ck. oil shales

Table 4.3 Range of Major Elements (weight%) in Julia Ck./Toolebuc Formation oil shales Bore hole

C	H	O	N	S	Ash	H/C	O/C
Tambo-38							
1–1	14	2.4	10.45	0.36	11.7	37	n.a
10–1	16	2.7	10.44	0.43	5.2	55.8	n.a
11–2	57.2	6.1	13.16	2.16	5.8	19.33	1.27
11–3	56.7	5.8	13.7	2.26	5.6	18.2	1.90
10–3	11.1	2.2	9.4	0.31	2.7	27.24	n.a
11–4	38.72	3.8	14.5	0.82	11.1	37.68	1.16
2–1	21.45	2.2	7.57	0.55	28.8	53.32	1.22
6–2	10.58	2.0	7.37	0.2	6.3	67.41	n.a
2.2	21.8	2.4	11.2	0.5	22.2	43.47	1.29
9–3	23.9	3.0	14.57	0.44	6.5	39.0	n.a
2–3	28.68	3.6	13.5	1.13	4.2	44.46	n.a
6–3	57.14	6.0	13.3	2.0	3.4	17.04	1.25
1–3	52.5	5.7	13.7	2.0	2.7	20.78	1.29
10–2	58.23	6.3	10.0	2.5	4.0	15.5	1.30
11–1	59.6	6.3	11.9	1.7	5.6	14.6	1.25
Longreach-6							
L6–1	44.1	5.4	12.7	1.4	5.7	43.8	1.37
L6–2	49.4	5.6	13.5	1.7	6.9	26.7	1.37
L6–3	38.7	4.9	13.7	1.5	5.4	38.0	n.a
L6–4	53.27	6.0	14.8	2.1	6.0	42.5	1.35
L6–5	54.4	6.0	10.1	2.0	5.3	23.6	1.33
L6–6	51.2	5.6	13.3	1.6	4.2	33.6	1.31
L6–8	63.38	6.5	9.5	2.0	6.3	14.6	1.30
L6–9	59.0	6.3	12.6	1.5	4.2	21.0	1.27
L6–7	56.38	5.7	8.3	2.0	8.0	26.5	1.21
L6–10	61.0	6.2	11.7	1.7	6.6	18.1	1.21
L6–11	53.8	5.6	13.0	1.6	6.7	25.8	1.24
L6–12	22.6	3.6	10.8	0.9	3.9	35.2	n.a
Jericho-11							
J11–2	9.3	2.0	9.2	0.4	5.3	43.0	n.a
J11–8	11.6	2.1	8.7	0.6	3.1	48.3	n.a
J11–9	8.9	1.7	5.4	0.2	5.3	54;1	n.a
J11–11	55.3	5.7	9.3	0	4.8	22.9	1.25
J11–12	8.2	1.9	15.0	0.4	3.2	39.2	n.a

(continued)

Table 4.3 (continued)

C	H	O	N	S	Ash	H/C	O/C
J11–24	45.0	5.3	14.5	1.7	6.2	42.0	n.a
J11–16	47.0	5.0	14.0	1.4	0.6	32.8	n.a
J11–18	54.3	5.8	12.6	2.0	3.5	23.5	1.21
J11–19	59.5	6.0	13.0	2.4	3.8	15.8	1.24
J11–25	32.7	3.9	16.0	1.1	4.0	28.0	n.a
J11–20	20.0	2.4	15.4	0.6	1.0	31.2	n.a
J11–21	13.8	2.0	15.6	0.4	2.9	33.4	n.a
Julia							
Creek	66.9	7.0	8.1	1.6	8.9	2.6	1.65
Boulia-3							
B3–1	56.3	6.1	13.0	1.7	6.6	13.6	1.30
B3–3	50.0	5.4	6.6	1.9	11.3	26.0	1.29
B3–4	27.0	3.7	15.8	0.8	4.5	37.6	n.a
B3–5	53.1	5.6	11.6	2.0	7.8	16.0	1.26
B3–6	47.0	5.3	10.9	0	22.6	26.1	n.a

n.a. = not applicable C% too low

containing highly aliphatic hydrocarbons which in turn reflect on an algal origin of this type of OM, which is low in oxygen and is highly desirable as an oil source. Torbanites, Kukersite, Tasmanite and Green River oil shales are Type I kerogen.

As clearly observed from the H/C ratios of the Julia Ck./Toolebuc Fm. oil shale samples, with the possible exception of Julia Ck. and Boulia boreholes all fit in Type II OM (kerogen) or even Type III OM. Julia Creek (JC) and Boulia boreholes contain samples that may be regarded as oil shales in the sense of commercial level of oil production, although a borderline case. Most samples had very low organic carbon to be regarded as oil shales. 50% organic carbon was used in the present study as the cut-off for H/C atomic ratios which would render them meaningless.

Depositional Environment of the Julia Ck./Toolebuc Formation

During the Lower Cretaceous Australia extended between latitudes 40° and 70° (Frakes 1979) with a warm current along its east shore which includes the Eromanga Basin (including the Toolebuc oil shales and evaporites along the eastern margins). The Cretaceous epoch is known for the absence of glaciation and ice caps. An average temperature of 32 °C has been suggested for equatorial surface waters in the west Pacific during the Albian, and an average temperature of 19 °C for polar regions in the west Pacific (Frakes 1979). During the Albian the Eromanga Basin was positioned between 45° and 50° latitude. The surface temperatures of the sea were in the range between 20 and 28 °C. A maximum temperature of 17 °C was put forward for Turonian seas in Australia by Stevens and Clayton (1971). The palaeo-temperatures of about 15 °C for bottom waters in the equatorial Pacific were calculated by Douglas and Savin (1975) from benthic foraminifera. The vast quantities of coccoliths remains point to optimum conditions for their prolific abundance. The optimum conditions for the growth of the common marine coccolith *Coccolithus huxleyii* is in the 18–24 °C range and salinities exceeding 34‰. The precipitation of carbonate within the cyanobacterial mats indicates high pH (above 7). Salinities associated with such mats at present are higher than those of average sea water, i.e. hypersaline in Solar Lake (Krumbein et al. 1977), 50‰ in Shark Bay and 40–50‰ in Spencer Gulf (Burne et al. 1980). Present-day water temperature for Spencer Gulf is between 12 and 25 °C, and for Shark Bay 18 and 24 °C are in the range of today's habitat of cyanobacterial mats. Denser bottom waters would be expected to have higher salinities.

Undisturbed shells in most Toolebuc Formation horizons indicate a quiet water environment. Stagnant conditions of bottom waters are also evident from concentrations of OM and particularly the high content of pyrite. Filamentous remains of inferred cyanobacteria in the OM of the Toolebuc oil shales occasionally show septate structure closely resembling the anoxygenic photosynthesising *Oscillatoria.* The rich plankton in the upper water column would have brought about oxygen depletion in the lower part of the water column. A close association of *Oscillatoria* with sulphate reducing bacteria is well documented.

The high nitrogen content in the Toolebuc oil shales organic matter presents yet another evidence for cyanobacterial activity. Nitrogen-fixing cyanobacteria are expected to be responsible for the relatively high Nitrogen concentration in these oil shales which are poor in humic acids. Humic acids are usually considered responsible for high N concentrations. Nitrogen shows close positive correlation with Carbon (Fig. 4.17).

Cyanobacteria are relatively low in lipidic content; 4–9% dry weight and high in carbohydrates and proteins (Wolk 1973) in contrast to green algae such as *Botryococcus* with lipid content of 30–40% dry weight reported (Wake and Hillen 1981).

Sulphur and Pyrite

The main source of sulphur in water is dissolved SO4, as organic sulphur from decomposition of proteinaceous matter is insufficient to contribute in any significant manner to the sulphate reservoir (Berner 1971). Thus, the amount of H_2S and consequently pyrite present in the sediment is the result of the amount of available sulphate, which in sea water is unlimited, and the availability of metabolizable organic matter for bacterial sulphate reduction. Sulphate reducing bacteria derive energy from oxidation of organic carbon anaerobically. It has been argued (Taylor et al. 1998) that pyrite can only form by bacterial activity, since on the basis of reaction kinetics there is insufficient energy for a purely chemical reduction of sulphates to disulphides. Toolebuc Formation oil shales are particularly high in pyrite seen preserved within the shells of foraminifera as well as within OM. When plotting C and S in the Toolebuc samples, a direct correlation is evident in the Boulia and Julia Ck. samples; whereas lesser positive correlation is evident in samples low in carbon and low H/C.

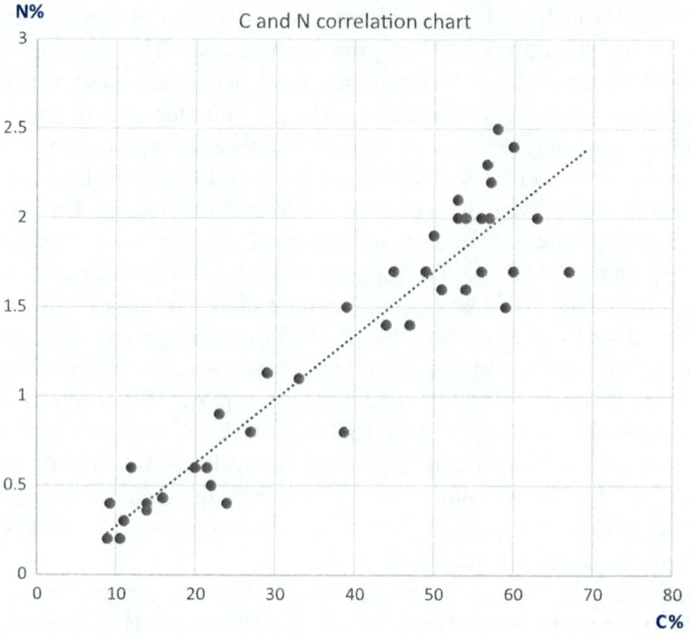

Fig. 4.17 Relation between C and N in Toolebuc Formation

The relatively low humic acid concentration in the Toolebuc samples is notable. Possibly the result of high bacterial activity. The latter used up the greater part of amino acids and carbohydrates and only a small fraction of these substances was left to form humic acids through melanoidin reactions. Most of the humic acids in the Toolebuc shales are *insitu* humates as shown from their isotope compositions; only a few samples indicated an allochthonous source.

4.1.4 Trace Elements

The association of trace metals with organic matter raises questions regarding the mode of concentrations, whether in the organic fractions or the mineral matter, and whether particular organisms have a role in the concentration of particular metals. The affinity and relationship of particular groups of organisms to trace metals has been the subject of numerous studies. Brown and Lester (1979) demonstrated the role of bacteria in flocculation and subsequently the precipitation of metals in the OM. According to these studies, the metals may be removed by absorption onto sites on bacterial extra-cellular polymers, to bacterial cell walls or accumulated in the cytoplasm of bacterial cells. TEM of the present study clearly shows absorption of metals in some bacteria on to their outer wall making them highly electron dense.

This is particularly noticeable in coccoid forms of bacteria. Some of the preservation of OM in the geological record may be explained as a result of natural metal staining.

Studies carried out on the uptake and retention of metals in *Bacillus subtilis* (Beveridge and Murray 1976) combined TEM and atomic absorption methods to demonstrate that there were preferred sites for the absorption and retention of various cations; namely, that Mg, Fe, Cu, Na K were deposited in the wall of the bacterium in large quantities, whereas Mn, Zn and Ca were retained in intermediate amounts, and only small amounts Hg, Sr, Pb and Ag. The elements Li, Ba, Co, and Al were not taken into the wall. Elements of higher atomic number could be detected as electron dense particles, diffusely staining the walls of the bacterium. Similar results are shown in the present study for metal absorption by humic acids. Humic acids absorb metals while lipidic materials such as green algae and some bacteria do not sbsorb metals as demonstrated from present study experiments (Fig. 4.24).

The Toolebuc oil shales being relatively organic-rich sediments provided an absorptive medium for various trace elements. Ni, V, Mo, and P have higher concentrations than average in these oil shales. Studies of trace elements in the Toolebuc oil shales were mostly concerned with vanadium, which occurs in unusually high concentrations in these oil shales. Riley and Saxby (1983) have shown that vanadium in the Toolebuc occurs mainly in three forms: (1) As vanadates or hydrated oxides absorbed onto clays and limestone; (2) in the form of vanadium silicates as part of clays-mica structure; and (3) as vanadyl porphyrins (Table 4.4).

Ekstrom et al. (1982) who noted, that nickel and vanadium porphyrins are the major metal–organic complexes present in the Julia Ck. deposits, also suggested that the Nickel and Vanadium have different origins since they are complexed to different types of porphyrins. As nickel is present in significantly smaller concentrations than vanadium in the Toolebuc sediments, vanadyl complexing was naturally more abundant. The question posed is what are the sources of the two kinds of porphyrins; whether they originate from different organisms, and if so what organisms. The other question raised is whether all the vanadium has its origin in the organic matter and was later mobilised into the inorganic fraction, as suggested by Riley and Saxby (1983), and if so, which organisms were responsible for such high concentrations of this element.

Analyses carried out in the present study as well as earlier studies (Dale and Fardy 1983) show a negative correlation between phosphorus and vanadium. Phosphorus and vanadium are known to compete for binding sites on tunicates. Tunicates were abundant organisms in the Cretaceous sea, and they are the only known organisms to concentrate high levels of vanadium from sea water (Wedepohl 1979; Macara et al. 1979; Lebar et al. 2011) and have up to 10% concentrations of this element in their blood. Vanadium concentration in the Toolebuc is up to 1%! The Julia Ck. oil shales are considered to have the world's highest V concentration and are presently being lobbied by a company as a source of Vanadium to be mined.

Vanadium in the Toolebuc Formation oil shales was most probably initially concentrated by the tunicates. Calcite spicules attributable to tunicates are clearly visible in thin sections of the oil shales and as white specks in reflected light of polished rocks (Figs. 4.4 and 4.5a). As TEM of the OM concentrate is after acid

Table 4.4 Concentration of Vanadium (V) in Toolebuc oil shales

Sample ID	V in whole rock ppm	V in OM ppm
1–2	890	180
1–3	620	n.d
B3-3	580	n.d
B3-4	1450	n.d
6–3	1450	330
B3-5	1100	n.d
10–2	630	69
10–3	430	42
11–1	1040	260
J11–24	545	n.d
11–2	720	54
11–3	890	–
L6–1	1170	170
L6–2	1220	230
L6–7	1380	–
L6–11	1660	–
L6–12	270	–
J11–11	680	n.d
J11–16	412	290
L6–10	1300	n.d

Below level of detection (–); n.d. not determined

solution of minerals, spicules would have been dissolved and are therefore no longer detected in TEM.

Following the death of tunicates, the vanadium was reduced by bacteria such as *Micrococcus lactilyticus* in the anoxic zone (where the OM was concentrated). Vanadium was later leached out of the organic matter and deposited within the inorganic fraction. The blood and soft parts of the tunicates were prone to bio-degradation. In the present study, a comparison between the vanadium content and organic carbon (Fig. 4.18, 4.19) shows a close correlation; vanadium mostly follows C, indicating that generally the greater part of the vanadium is still contained within the organic fraction. Similar results were obtained from a study isolating Vanadyl and Nickel pophyrins from a Cretaceous oil shale in Morocco (Saoiabai et al. 1983).

Vanadium in the OM fraction is in the form of vanadyl porphyrins (Riley and Saxby 1983; Ekstrom et al. 1982). It is suggested in the present study that the vanadium porphyrins are the original vanadium pigments of ascidians which are known to have possessed vanadium pigments in *vanodocytes*, their specialised blood cells, as well as other forms of vanadium such as V_2O5 present in the rest of the cells of these organisms. The vanadium concentrated by these organisms in other than pigment form would have most probably been released following their decomposition, and

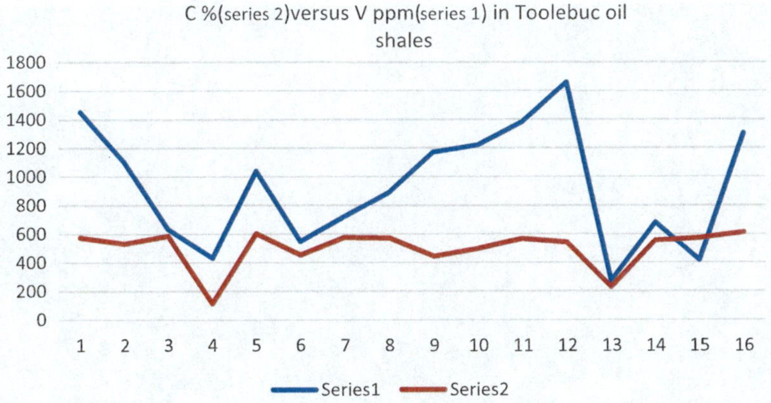

Fig. 4.18 Relation between carbon and vanadium in Julia Ck. oil shales

mobilised, with the consequent precipitation and deposition in the inorganic fraction. The spicules of tunicates visible in polished rock samples (Figs. 4.4; 4.5) are evidence of their high concentrations in these environments.

Although Mo shows some positive correlation with C content, it is mainly concentrated in the inorganic whole rock fraction, which may have had its source in the organic fraction, and migration into the rock following degradation of the organisms that concentrated the element initially.

In addition to the high vanadium concentrations, high Mo and Zn levels, moderate Cu and Ni concentrations obtained in the present study (Table 4.5) have also been reported with a relative abundance of rare earths elements in previous studies (Dale and Fardy 1983).

Nickel, when present in significant quantities, originates from ultrabasic and basic rocks: Weathering of these rocks contributes to high concentrations of this metal in sea water.

The concentration of Ni in sea water is 0.6 ppm in the black sea, and varies from 0.01 to 0.3 in other sea water. Brumsack (1980) found Ni, Cu, Zn and Fe to be associated with sulphides and hence, normal sea water is sufficient to account for the high metal concentrations in black shales, in spite of a definite correlation of the metals with organic carbon. Brumsack (1980) explained the association of heavy metals with organic matter as indirect, namely the anoxic basin acting as a sink for heavy metals due to the high sulphide content. This May account for the presence of a certain part of these metals in the black shales. However, nickel may also be present in porphyrins, where it replaces Mg in the chlorophyll molecule, and may therefore be directly associated with organic carbon content. In the present study, Nickel is found to be associated with the organic as well as inorganic fractions with no apparent preference towards the one or the other. No direct overall linear relationship is evident between nickel and organic carbon, but some positive correlation is evident in a few samples where Ni clearly follows C (Fig. 4.21). Ni concentrations in the Toolebuc oil shale samples in this study vary from 60 to 274 ppm in whole rock

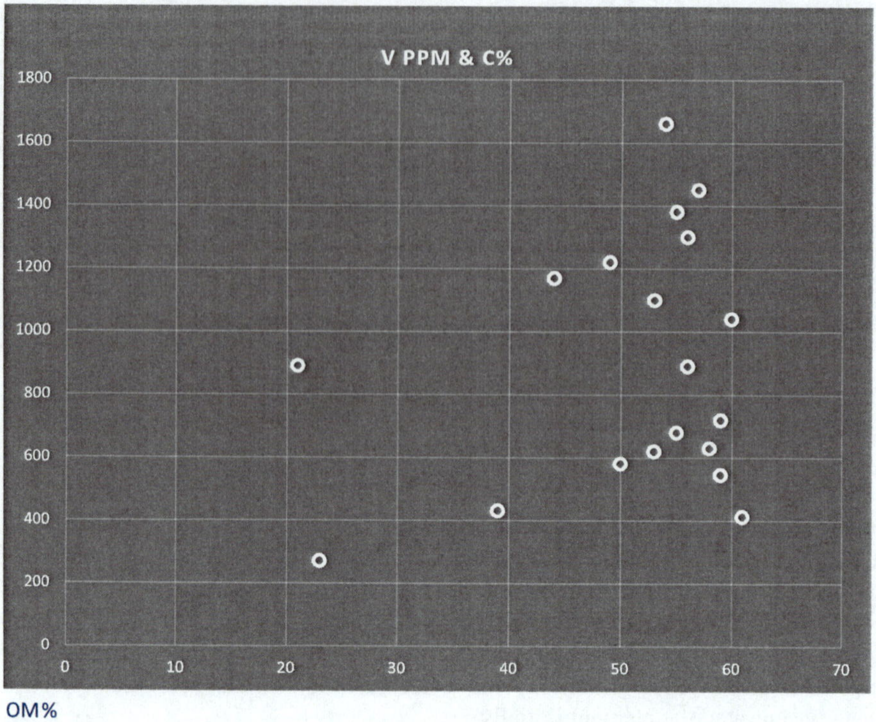

Fig. 4.19 V (ppm) concentrations

samples and 41–516 ppm in the organic matter concentrate after demineralisation. By comparison, a sample from the underlying Corinna sandstone yielded 24–10 ppm in whole rock and OM samples respectively. Certainly, a significant reduction of Ni compared to the oil shale facies.

The dependence of nitrogen-fixing cyanobacteria such as *Oscillatoria* on the presence of Mo for the biosynthesis and activity of the enzyme nitrogenase has been demonstrated by Cardenas and Mortenson (1975). Sullivan et al. (1980) have shown that Mo and Vanadium inhibit the oxidation of sulphur compounds by the *Thiobacilli*. This finding would have a bearing on the optimum conditions for cyanobacterial development in regard to anoxygenic photosynthesis (Glikson and Taylor 1985).

Some heavy metals are precipitated directly as sulphides as argued by Brumsack (1980). However, many metals are known to be concentrated by organic matter either directly by live organisms or indirectly by dead algal or other microorganisms' remains in the form of particulate organic matter. Nickel concentrations in the Toolebuc Formation do not exceed those reported by Brumsack (1980) in North Atlantic black shales and the hypothesis of direct precipitation as sulphides may be valid, although only a weak relationship between S and Ni is evident in the Toolebuc oil shales.

Table 4.5 Nickel concentrations in the Toolebuc oil shales

Sample	In whole rock ppm	In organic concentrate ppm
1–2	135	168
6–3	207	–
10–2	81	54
10–3	70	39
11–1	182	140
11–2	139	41
11–3	159	–
L6–1	142	206
L6–2	145	312
L6–7	261	–
L6–11	274	–
L6–12	89	211
L6–6	–	87
L6–10	231	516
L6–9	185	n.d
B3–3	89	n.d
B3–4	144	n.d
B3–5	260	n.d
J11–24	172	n.d
J11–11	199	n.d
J11–16	237	n.d

Below level of detection (–), n.d. = not determined

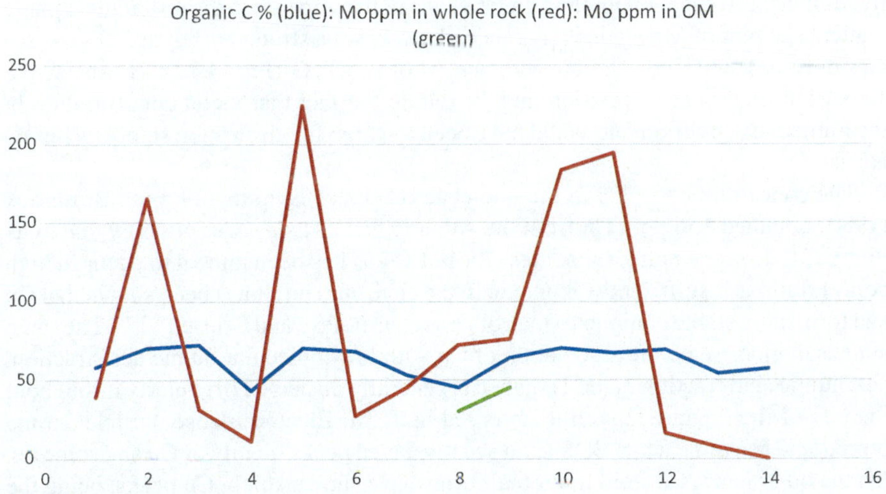

Fig. 4.20 Molybdenum in Toolebuc oil shales

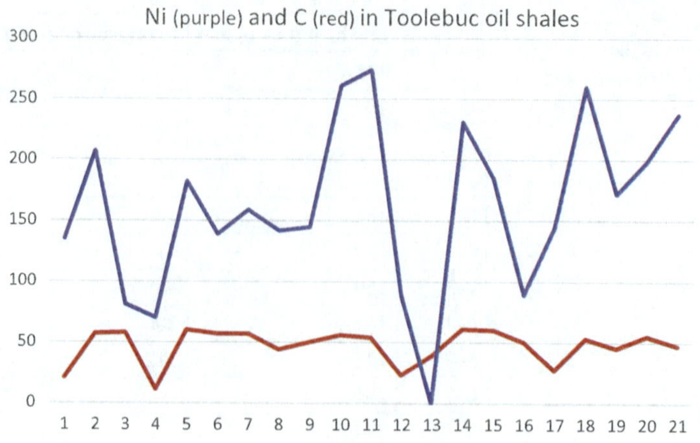

Fig. 4.21 C%, Ni ppm

Pilipchuk and Volkov (1974) in their studies of Mo behaviour in the Black Sea found that enrichment of Mo in the anoxic zone was the result of two main processes, namely initially diagenetic fixation of the metal to organic matter followed by precipitation with pyrite. A relationship between Mo and sulphur has been noted in the present study.

It has been observed in studies of metal interactions in recent sediments (Nissenbaum and Swaine 1976) that marine humic substances concentrate Cu, Mo and Zn in large proportions. Humic acids generally have been shown to be less important in the concentration of Ni, Co, and Pb. These conclusions were attributed to the presence of specific ligands in marine humates. In the present study, the humic acid fraction in the Toolebuc samples is relatively small compared to the rest of the organic matter. The plot of Mo against organic carbon shows a strong correlation (Fig. 4.20) and only moderate correlation with the whole rock C (Fig. 4.22). However, the somewhat imperfect correlation may be due to the fact that metal concentration in the immediate environment would have been too high for absorption solely by humic acids.

The concentration of Cu in the Toolebuc oil shales is up to 129 ppm. Brumsack (1980) obtained 156 ppm Cu from the Atlantic black shales. Cu, possibly due to its unusually strong bonding capacities (Rashid 1974) has been known to occur in high concentrations if sufficiently abundant in the environment. It has been shown that Cu can form more stable complexes than other metals (Giesy and Briese 1977). The same studies demonstrated the association of Cu with low molecular humic acid fraction. The humic acid fraction in the Toolebuc is generally dispersed irregularly throughout the rest of the organic matter as observed in TEM. Electron dense particles in the humic acid fractions when EDS analysed displayed peaks mainly of Cu and some Fe. On the other hand, Coccoid microbial forms have shown solely Cu peaks, being the metal causing high electron density of these microbes. As mentioned earlier metal absorption by bacteria has been studied extensively (Beveridge and Murray 1980)

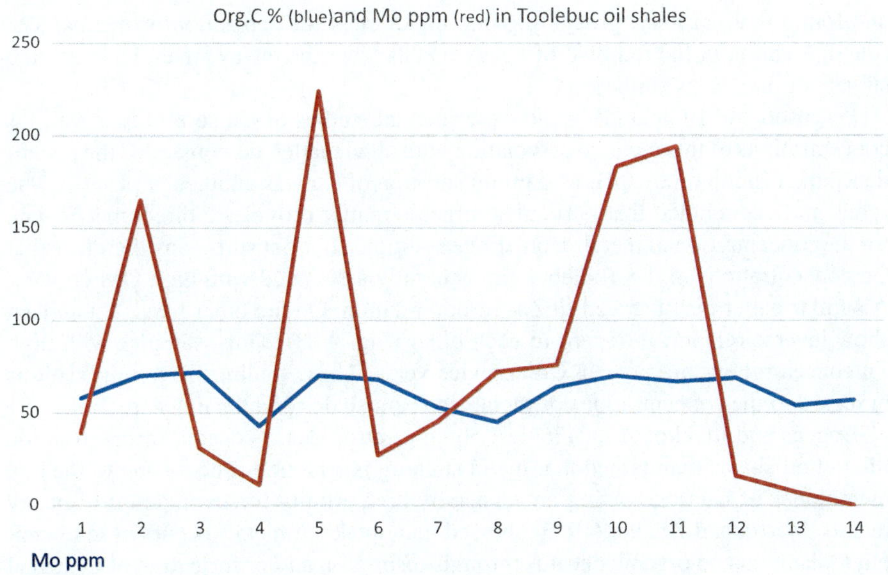

Fig. 4.22 Relation between carbon and molybdenum in Julia Ck. oil shales

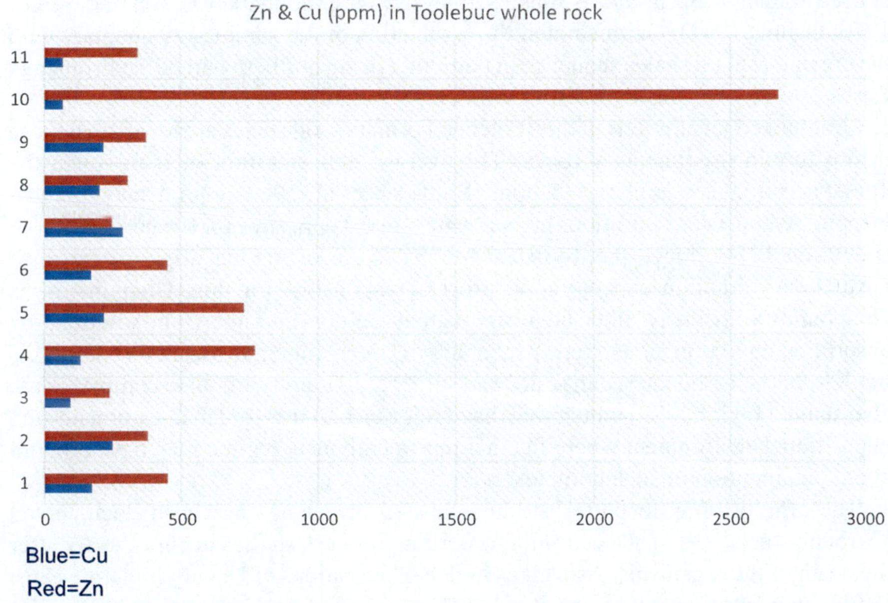

Fig. 4.23 Concentrations of Zn and Cu in Julia Ck. oil shales

and found that carboxyl groups provide major sites for deposition of metals. The role of bacteria in the removal of heavy metals from the environment has been the subject of numerous studies.

Ferguson and Bubela (1974) in experimental studies of Cu solubility at various concentrations of this metal in association with algal matter, demonstrated the dependence of sorption on the pH. Maximum sorption of Cu was attained at pH = 6. The same study concluded that particulate organic matter provides a direct mechanism for the concentration of metals from dilute solutions. It is not surprising therefore that Cu concentrations in the Toolebuc are generally associated with high OM content. A similar pattern is observed in Zn–carbon relation. On the other hand, Cu and Zn show inverse relation in regard to each other (Fig. 4.23). Thus, samples with high Zn concentrations are poor in Cu and vice versa. These findings have implications in regard to the concentrations of metals in spent shales (details in Chap. 7).

Degens and Ittekkot (1982) in describing electron dense coccoid forms from the Messel oil shales maintained that metal staining is a natural process due to the cell membranes of bacteria having an especially high affinity for the fixation of heavy metals. Martin and Knauer (1973) showed that uptake of metals occurs in senescent phytoplankton and organic detritus through absorption and/or formation of hydroxyl complexes on the surfaces of organic matter.

Uranium in the Toolebuc Formation is noted as a gamma-ray anomaly over the entire Eromanga Basin and is thus used as stratigraphic marker (Exon and Senior 1976; Ozimic 1982). Ozimic noted the association of the gamma-ray anomaly with the organic-rich oil shales sequence in particular in the northern part of the Eromanga Basin.

Uranium concentrations in the Toolebuc Formation oil shales are lower than those in ore forming sedimentary rocks. The average concentration of uranium in the Toolebuc oil shales is about 35 ppm (Ramsden et al. 1982) with a maximum of 80 ppm. Average concentrations in river water are 0.4 ppm and in sea water 0.002 ppm (Levinson 1974). Since the Toolebuc oil shales are of marine origin, there must have been a uranium concentration process responsible for this. Uranium forms compounds with many other elements, it may be concentrated from solutions by absorption on to organic matter. A secondary U deposition may occur as complexes of phosphates, carbonates, vanadates and sulphates. Uranyl carbonate complexes are also stable (Taylor 1979) which may have relevance to the Toolebuc carbonate-rich depositional environment where fluctuations in carbonate regime may have brought about precipitation of such complexes.

The affinity of U to phosphate and thus to fish bones has been documented (Arrhenius et al. 1957). The sorption according to these studies occurs shortly after the death of the organisms. As well as high concentrations of U, very high rare earth (REE) elements concentrations (0.6–1.5%) were found in association. Unusually high REE have been recorded in the Toolebuc oil shales (Dale and Fardy 1983). Similar associations between U3O8, P2O5 and organic carbon have been reported by Ramsden et al. (1982) in the Toolebuc Formation oil shales. These studies showed a positive correlation between organic carbon and uranium, and between phosphate and uranium. Where OM was present in high concentrations, U is seen to have been

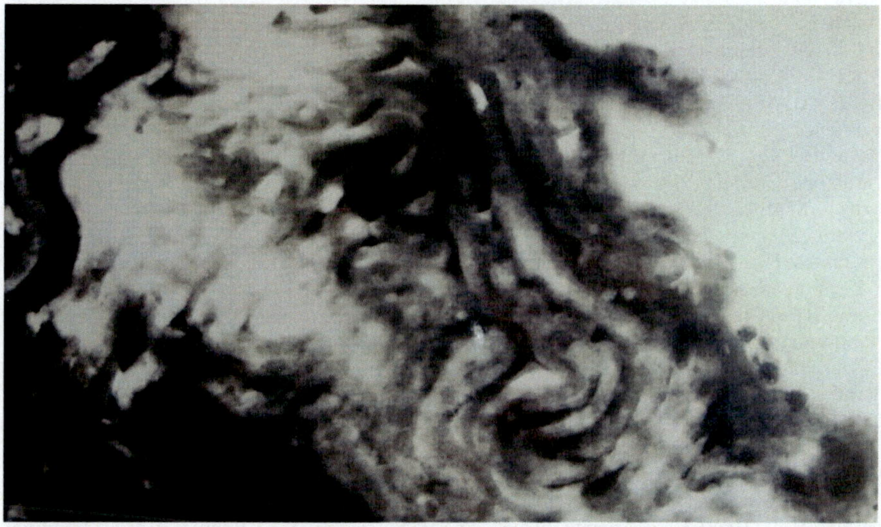

Fig. 4.24 TEM of microbial lipidic filaments. Metal stain applied was not absorbed by the microbial filaments, but was absorbed by degraded OM. X 40 000

absorbed by it. However, in limestones overlying the oil shale sequence, where phosphate concentrations are high, but organic carbon is low, the U is associated with the phosphate.

It has been demonstrated experimentally (Horikoshi et al. 1979) that U is concentrated by particulate organic matter. These authors used cultures of Chlorella live cells as well as scalded dead cells of the alga, and found that uptake was dependant on physical absorption of cell walls and not on the biological activity of the organism. The degree of sorption depended on pH. Maximum concentrations were attained at pH = 5 and higher (up to pH = 9) in scalded cells, whereas in living cells maximum sorption was limited to pH = 5–6 due to inhibitors of toxicity in the living alga. Scalded cells showed a much higher uptake than living cells. This supports humic acids, being the degradation products of OM as a focus for U sorption. Furthermore, this explains the high absorption of metals to the extent of total opacity in the coccoid bacteria (Figs. 4.14). On the other hand, lipidic filamentous microbes do not absorb metals (Fig. 4.24)

References

Arrhenius, G., Bramlette, M. N., & Picciotte, E. (1957). Localisation of radioactive and stable heavy nucleids in ocean sediments. *Nature, 180,* 85–86.

Bauld, J. (1980). Production and fate of organic carbon in blue-green algal mats. *Baas-Becking Geobiology Laboratoriya Annals Reports, 1980,* 19.

Bauld, J. (1981). Geobiological role of cyanobacterial mats in sedimentary environments: Production and preservation of organic matter. *BMR Journal Australian Geologico Geophysics, 6,* 307–317.
Berner, R. A. (1968). Calcium carbonate concretions formed by the decomposition of organic matter. *Science, 59,* 195–197.
Beveridge, T. J., & Murray, R. G. E. (1980). Sites of metal deposition in the cell wall of Bacillus subtilis. *Journal of Bacteriology, 141,* 876–887.
Brown, M. J., & Lester, J. N. (1979). Metal removal in activated sludge: The role of bacterial extra-cellular polymers. *Water Research, 13,* 817–837.
Brumsack, H. J. (1980). Geochemistry of cretaceous black shales from the Atlantic ocean (DSDP Legs 11, 14, 36, 41). *Chemistry Geologico, 31,* 1–25.
Burne, R. V., Colwell, J. B., Pain, L., & Tratt, M. M. (1980). Sedimentological studies in the spencer gulf. *Baas-Becking Geobiology Laboratoriya Annals Reports, 1980,* 11–16.
Calder, J. A., & Parker, P. L. (1973). Geochemical implications of induced changes in 13C fractionation by blue-green algae. *Geochimica Et Cosmochimica Acta, 37,* 133–140.
Cardenas, J., & Mortenson, L. E. (1975). Role of Molybdenum in nitrogen fixation by Clostridium pasteurianum. *Journal Bacteria, 122,* 978–984.
Dale, L., & Fardy, J. J. (1983). Trace element characterisation of oil shales. Proc.1st Austral. Workshop on oil shales: 57–60, Lukas Heights.
Degens, E. T. (1969). Biogeochemistry of stable carbon isotopes. In G. Eglinton & M. T. J. Murphy (Eds.), *Organic Geochemistry* (pp. 304–309). Berlin: Springer Verlag.
Degens, E. T., & Epstein, S. (1964). Oxygenated Carbon isotope ratio in coexisting calcites and dolomites. *Geochimica Et Cosmochimica Acta, 28,* 23–44.
Degens, E. T., & Ittekott, V. (1982). In-situ metal staining of biological membranes in Sediments. *Nature, 298,* 262–268.
Eadie, B. J., Jeffrey, L. M., & Sackett, W. M. (1978). Some observations on the stable isotope composition of dissolved and particulate organic carbon in the marine environenment. *Geochimica Et Cosmochimica Acta, 42,* 1265–1269.
Ekstrom, A., Loeh, H., & Dale, L. (1982). The petroporphyrins in the oil shales from the Julia Creek deposit. *Eromanga Basin Symposium Petrology Exploration Societa and Geological Societa Australia*, 215–216.
Exon, N., & Senior, B. R. (1976). The Cretaceous of Eromanga and Surat Basins. *BMR Journal of Australian Geology and Geophysics, 6,* 1–29.
Ferguson, J., & Bubela, B. (1974). The concentration of Cu, Pb, and Zn from aqueous solutions by particulate algal matter. *Chemical Geology, 13,* 163–186.
Frakes, L. A. (1979). *Climates Throughout Geologic Time.* Amsterdam: Elsevier.
Giesy, J. P., & Briese, L. A. (1977). Metal associated with organic carbon extracted from Okefenokee swamp water. *Chemical Geology, 20,* 109–120.
Glikson, M., & Taylor, G. H. (1985). Cyanobacterial mats: Major contributors to organic matter in Toolebuc Formation oil shales. *Geology Society Australia Sp. Bubl. No, 12,* 273–286.
Glikson-Simpson, M. (2020). *Coal-Window to Past Climate and Vegetation.* In print: Springer.
Horikoshi, T., Nakajima, A., & Sakaguchi, T. (1979). Uptake of uranium by chlorella regularis. *Agricultural Biology Chemistry, 43,* 617–623.
Hulbert, E. M., & Rodman, J. (1963). Distribution of phytoplankton with respect to salinity between the coast of southern New England and Bermuda. *Limnology and Oceanography, 8,* 263–269.
Irwin, H., Curtis, C., & Coleman, M. (1977). Isotopic evidence for source diagenetic carbonates formed during burial of organic-rich sediments. *Nature, 269,* 109–213.
Krumbein, W. E., Cohen, Y., & Shilo, M. (1977). Solar Lake (Sinai) 4. *Stromatolitic Cyanobacterial Mats. Limnology Oceanography, 22,* 635–656.
Lebar, M. D., Luttenton, L., McClintock, J. B., & Baker, B. J. (2011). Accumulation of V, Mn, Ni in Antarctic tunicates. *Polar Biology, 34*(4), 587–590.

Martinez, C., Baldini, R. L., & Gomez, S. L. (2006). A caulobacter cressentus extracytoplasmic function factor mediating the response to oxidative stress in stationary phase. *Journal Bacteria, 188*(5), 1835–1846.

Malone, T. C., Garside, C., HINES, K. C., & Roels, O. A. (1975). Nitrate uptake and growth of Chaetoceros sp. in large outdoor cultures. *Limnology and Oceanography, 20,* 9–19.

Martin, J. H., & Knauer, G. A. (1973). The elemental composition of plankton. *Geochimica Et Cosmochimica Acta, 37,* 1639–1653.

Nishimura, M. (1977). The geochemical significance in early sedimentation of peolipids obtained by saponification of lacustrine sediments. *Geochimica Cosmiochimica Acta, 41,* 1817–1823.

Nissenbaum, A., & Swaine, D. J. (1976). Organic matter—metal interactions in recent sediments: The role of humic substances. *Geochimica Et Cosmochimica Acta, 40,* 809–816.

Ozimic, S. (1982) *Stratigraphic drilling in the Cretaceous Toolebuc formation facies and in their lateral equivalents, Eromanga nad Carpentaria Basins.* Eromanga Basin Symposium Summary papers. Petroleum Exploration Society and Geology Society Australia, (pp. 229–231).

Ozimic, S. (1983) Oil shale methodology. An examination of the Toolebuc Formation and laterally contiguous time equivalent units., *Eromanga and Carpetaria Basins.* NERDDC Project 78/2616: 1–59.

Pardue, J. W., Scalan, R. S., van Baalen, C., & Parker, P. L. (1976). Maximum carbon fractionation in photosynthesis by blue-green algae and green algae. *Geochimica Et Cosmochimica Acta, 21,* 110–126.

Philipchuk, M. F. & Volkov, I. I. (1974). Behaviour of Molybdenum in processes of sediment formation and diagenesis in the Black Sea. In Degens, E. T. & Ross, D. A. (Eds.) *The black sea geology, chemistry and biology.* AAPG Mem. (Vol. 20, pp. 542–553).

Ramsden, A. R., Dickson, B. L., & Meakins, R. I. (1982). Origin and significance of the Toolebuc gamma-ray anomaly in parts of the Eromanga Basin. *Journal of the Geological Society Australia, 29,* 285–296.

Rashid, M. A. (1974). Absorption of metals on sedimentary and peat humic acids. *Chemical Geology, 13,* 115–123.

Riley, K. W., & Saxby, J. D. (1983) Association of organic matter and Vanadium oil shale from the Toolebuc Formation of the Eromanga Basin, Australia. *Chemistry Geology*

Ryan, W. B. F., & Cita, M. B. (1977). Ignorance concerning episodes of ocean-wide stagnation. *Marine Geology, 23,* 197–215.

Saoiabai, A., Ferhat, M., Barbe, J. M., & Guilard, R. (1983). Metal including vanadyl and nickel porphyrins in oil shales of Timahdit, Morocco. *Fuel, 62,* 963–965.

Senior, B. R., Exon, N. F., & Burger, D. (1975). The Cadna-Owie and Toolebuc Formations in the Eromanga Basin Qunsland. *Queensland Governance Mining Jornal, 76*(890), 44–455.

Stevens, G. R., & Clayton, R. N. (1971). Oxygen isotope studies of Jurassic and Cretaceous Belemnites and their biogeographic significance. N.Z. *Geology Geophysics, 14,* 829–998.

Taylor, G. H., Teichmuller, M., Davis, A., Diessel, C. F. K., Littke, R., & Robert, P. (1998). *Organic Petrology* (p. 704). Berlin: Gebruder Borntraeger.

Tissot, B. P., & Welte, D. H. (1978). *Petroleum Formation and Occurrence* (p. 699). Berlin: Springer.

Wake, L. V., & Hillen, L. W. (1981). Nature and hydrocarbon content of blooms of the alga *Botryococcus braunii* occurring in Australian freshwater lakes. *Australian Journal Marine Freshwater Research, 32,* 353–367.

Wedepohl, K. H. (1979). Composition and abundance of common sedimentary rocks. In K. H. Wedepohl (Ed.), *Handbook of Geochemistry.* Berlin: Springer.

Wolk, C. P. (1973). Physiology and cytological chemistry of blue-green algae. *Bacteria Reviews, 37,* 32–101.

Chapter 5
Tertiary Oil Shales in Australia: Rundle Complex

Abstract Deposits of organic rich lacustrine sediments of early mid Eocene age have been preserved in elongate basins within a series of asymmetric grabens extending over 500 km along the central coastal region of Queensland. These grabens developed through a tensional regime following the opening of the Coral Sea in the early Tertiary (Veevers et al. 1991), in view of seafloor spreading around Australia at the time. A lake system developed over time with climatic fluctuations and structural movements bringing about changes in water level, water quality partial drying over time and complete drying, and renewal of the cycle with another wet regime. These changes were expressed in typical cyclicity of the sediments. Although a number of grabens and semi grabens of this part of the central coast contain oil shales the name 'Rundle' will be used as a general term for all the oil shale bearing deposits as they are of the same age and same depositional environment. Some of these deposits were mined and processed in the eighties with active research ongoing long after. The mining and processing of these oil shales is uneconomical and of potentially enormous environmental destruction in a particularly sensitive area, to farming, fishing, local rural population and the Great Barrier Reef Marine Park. A moratorium until 2028 has been declared on the mining of these oil shales by the Queensland government.

5.1 'Rundle' Complex Oil Shales

The Rundle Complex oil shales (Figure 5.1) which include the Narrows graben that contains the Rundle deposit in the strict sense and the Stuart oil shales within the 400 m deep Rundle Formation which is made up of cyclic depositional sediments of alternating oil shales, lignitic/coaly beds and claystones that have been logged in detail (Coshell 1983). The Hillsborough graben contains the oil shale deposits of Condor and Yamba. The Hillsborough graben offshore part lies entirely within the Great Barrier Reef Marine Park. The onshore part of the Hillsborough Graben has been researched as a target for oil production. The Duaringa Basin is within a semi graben further inland, and has been evaluated for commercial oil production that found the oil shales of poor to medium grade and high in moisture (Moore et al. 1987) and therefore would require high energy input for drying. The Condor

M. Glikson-Simpson, *Oil Shales*, https://doi.org/10.1007/978-3-030-60675-6_5

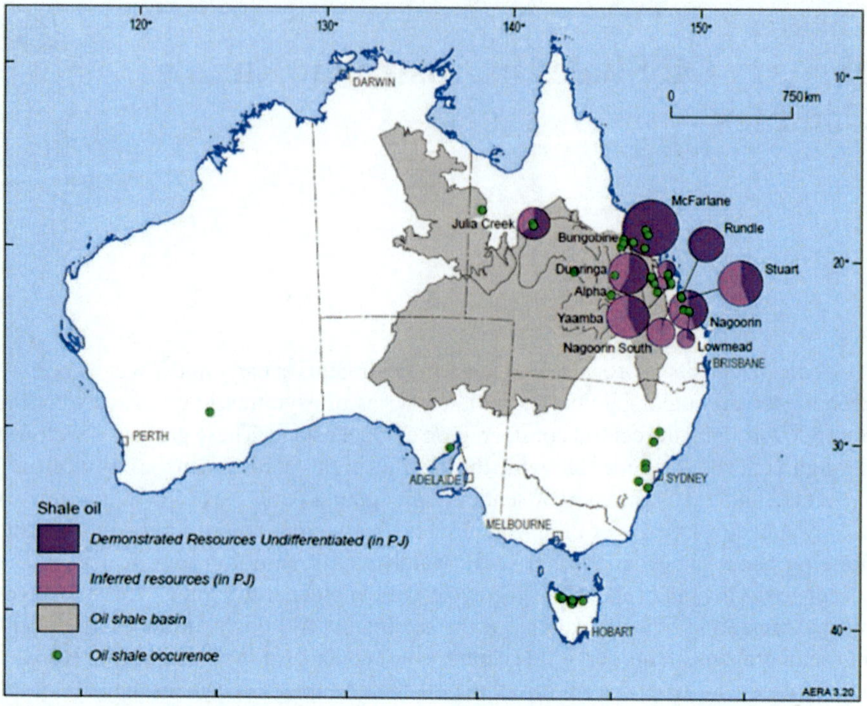

Fig. 5.1 Map of 'Rundle Complex' oil shale deposits. *Source* Geoscience Australia

deposit within the Hillsborough Graben had been researched extensively and found to be dominated by lignitic-coaly beds. Nagoorin Graben in the south of the coastal strip contains a cyclic sequence of over 400 m thick that contains some oil shales, lignite beds alternating with cannel coal (Henstridge and Hutton 1987). The Nagoorin graben has been divided into south and north. The southern part with the cyclicity of algal–dominated oil shale and cannel coal, Lignitic–coaly beds and intervals of mudstones.

The 'Rundle complex' oil shales which contain all the above-mentioned deposits within the Eocene grabens have yields of around 100 L/ton on average. While the oil shales proper may yield up to 200 l/ton, the finely inter-bedded lignites yield 50 l/ton or less. When mined, the inter-bedded lignite-dominated sediments are processed together with the oil shale beds. Thus, an average of about 70–80 l/ton is realistic. Most of the above- mentioned deposits are predominantly lignite beds, whereas only the Rundle and Stuart deposits are dominated by oil shale facies.

Intensive exploration by companies during the eighties led to pilot production, and retorts were constructed specially for oil production from some of these Tertiary oil shales along the coast of central Queensland. Attempts were made to demonstrate various rehabilitation procedures of the disturbed land, the enormous quantities of spent shale and the leaching of toxic elements. These will be discussed in greater detail in Chap. 7. An EIS stage 2 drilling project was completed in 1998. An interest in these oil shales continued on, though informally, in spite of the venture being totally uneconomical. The latest Resource Estimate Update has been published in 2014.

The term Rundle oil shales is used here generally, when not specified otherwise for all the deposits within the grabens and semi grabens along the central coast of Queensland. These oil shales are Eocene lacustrine deposits consisting of cyclic depositional sediments of oil shale seams inter-bedded with non-oil shale beds of lignitic-coaly material. The thickness of the oil shale containing formations varies in the different grabens but is generally between 400 and 500 m thick. The Stuart oil shale deposit lies in the southern half of the Narrows graben while the Rundle oil shales lie within the northern part of the Narrows graben.

Although the entire area of Eocene depositional basins experienced the same climatic fluctuations, the differences in facies between the basins in terms of the composition of the cyclic sediments are evident. High input from swamp vegetation and formation of lignite beds are common in the Hillsborough graben Condor deposit, whereas algal dominated lake environment is evident in the Stuart-Rundle deposits of the Narrows graben. The climate fluctuations and possibly tectonic activity were responsible for flooding, water level fluctuations leading to fresh water versus brackish, and altogether drying periods represented by intervals of clastic sediments, ultimately leading to the differences between the various Rundle Complex deposits.

The organic matter of the Rundle oil shales is predominantly the alga Pediastrum (Fig. 5.2). In TEM, two types of dominant organic entities can be distinguished: (1) The predominant lipidic algal and (2) humic acids filling spaces between the algal remains (TEM Figs. 5.3a, b). The microstructure of the *Pediastrum* algal colonies is represented by stringers extending from the algal cells. Humic acids filling spaces

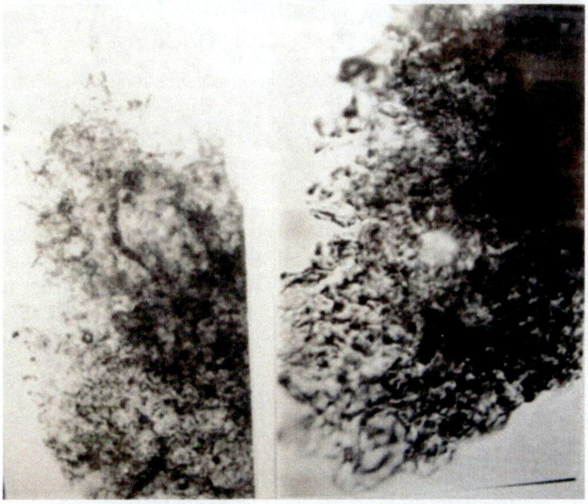

Fig. 5.2 Transmitted light microscopy of the *Pediastrum* algal colonies dominating the rundle complex OM. Photo taken at X250 mag

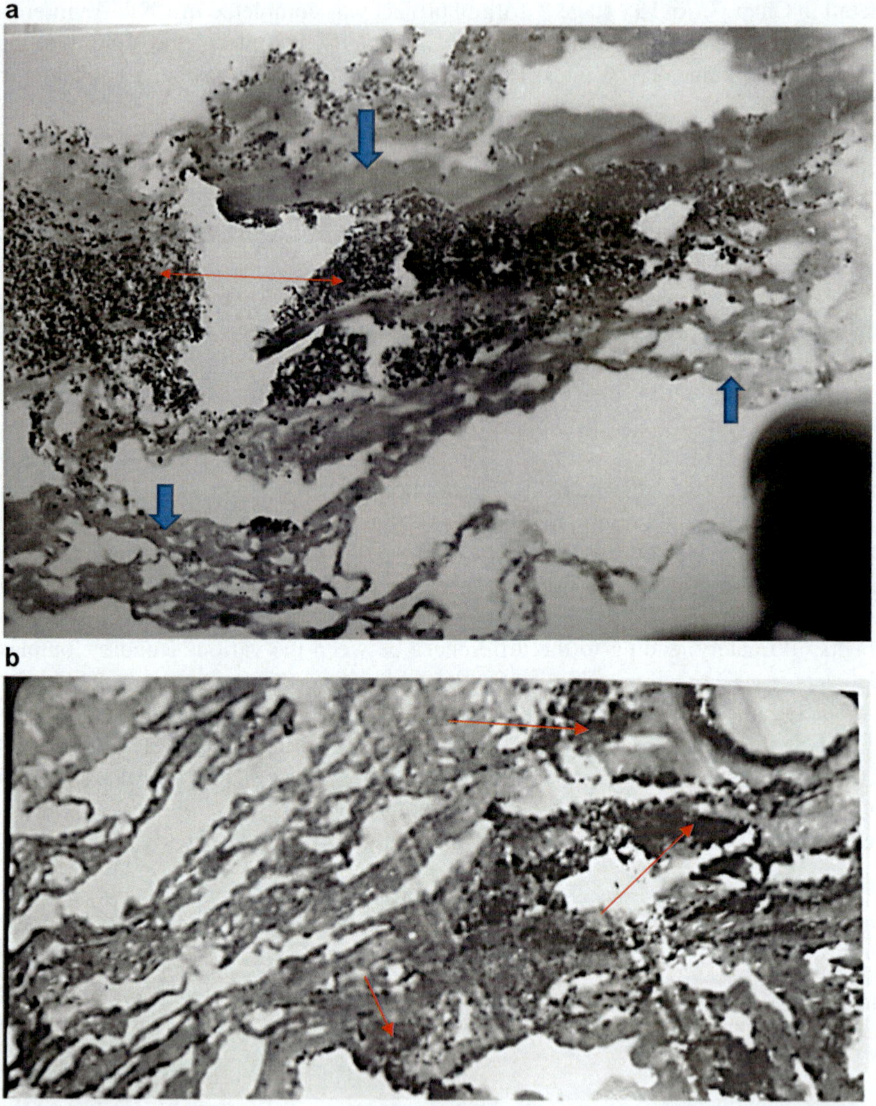

Fig. 5.3 **a** The two main components of the OM in Rundle Complex as seen in TEM: Algal material (thick blue arrow) and humic acids (red arrow) with metal concentrations (opaque particles). X 22 000. Note characteristic stringers of the green alga *Pediastrum.* **b** Detail of algal cell walls and metal concentrations in Rundle Complex. Arrow points to metals absorbed on to humic acids. Mag.X 22 000

Table 5.1 Major elements in Rundle-Condor oil shales compared with Tasmanite and Green River oil shales

Major Elements %:	C	H	O	N	S	Ash	H/C
Condor							
Con-1	54.81	6.85	6.58	1.76	2.10	21.58	1.49
Con-2	68.47	5.71	10.01	1.67	0.00	2.49	0.99
Rundle							
R-1	59.82	8.19	16.57	0.70	0.81	14.30	1.63
Tasmanite	72.97	9.52	9.40	0.81	8.91	6.69	1.55
Green river	68.57	8.79	8.94	0.93	3.54	8.70	1.53

between the algae may constitute up to 30% or more of the total OM, detracting from the oil potential of the deposits. Electron opaque particles absorbed on to the humic acids are evident. EDS analysis showed Ti the most common mineral. Dale and Fardy (1983) reported 0.4% Ti in trace element analysis of Rundle oil shales. Trace element analysis in the present study showed close to 6000 ppm Ti in the Rundle Complex oil shales, compared to 6500 ppm in Green River.

Humic acids are particularly powerful as solvents of minerals (Baker 1973) in relatively high rainfall climates and high intensity of weathering. Cheshire et al. (1977) found that Ti was concentrated mainly by humic acids. The volcanic rocks associated with the Rundle Complex oil shales acted as a source for the high concentrations of Ti in these deposits.

Major elements in the Rundle complex oil shales compare well with Green River and Tasmanite. The H/C ratio however drops in lignite beds like Condor-2 where high oxygen is evident. Baker (1973) demonstrated the role that humic acids play in mineral degradation and solubilisation. In particular Ca, Pb, Cu and Ni have been shown to be highly soluble in humic acids. Sulphur is generally low as evident in elemental analyses of the organic matter. It is mostly tied up in pyrite as indicated by EDS of TEM–observed opaque particles. Therefore, the iron sulphide will be represented in the ash content of the spent shale. Lacustrine organic-rich deposits are generally of low sulphur content compared to marine deposits due to low sulphate concentration compared to sea water and marine environments.

The algal domination in The Rundle-Stuart deposits within the Narrows Graben points to long periods of high-water level in the lake system. In contrast, the absence of algal accumulations in Condor supports a swamp environment in the Hillsborough graben.

Patterson et al. (1987) found above normal abundance of arsenic and manganese associated with pyritic sulphur in the Rundle oil shales. The present study obtained very high concentrations of Arsenic and other toxic metals in spent shale (details in Chap. 7).

The Condor–Rundle organic matter yielded carbon isotopic composition typical of planktonic organisms which are predominantly the components of the 'brown' parts of the oil shale sequence. However, the lignite black beds within the sequence

Table 5.2 C-isotope compositions of Rundle Complex OM

C–isotope composition:		$\delta^{13}C$ ppt
Rundle	OM less HA	−21.1
Condor brown	" " "	−22.2
" "	HA	−26.7
" black	OM less HA	−24.8

OM = Organic Matter
HA = Humic Acids

yielded isotopic values normally obtained from coals. Furthermore, the isotopic values obtained from the humic acid fraction in the lignite-dominated Condor deposit are significantly lighter than the algal material and more characteristic of coals and support a swamp vegetation as a source for these humates.

The Nagoorin graben oil shale deposit was detailed by Henstridge and Hutton (1987). The sequence is over 400 m thick and contains carbonaceous oil shales composed of alternating oil shales, stratified lignite beds interbedded with thin layers of cannel coals. In terms of depositional environment at the time it was a cyclic succession of swamp-dominated environment, flooded at times and back to swamp again. Oil yield reported for Nagoorin carbonaceous shale was about 80 l/ton (Noon 1987). Furthermore, on the basis of the high input of lignite which termed the Nagoorin 'deposit' carbonaceous Shale, it was deemed unsuitable for oil production as it would require "severe hydrotreatment, consuming large quantities of hydrogen" (Moore et al. 1987) and energy input.

Stuart oil shale deposit which lies within the southern half of the Narrows Graben has shown average oil yield of 134 l/ton (Moore et al. 1987). This oil shale deposit was regarded as commercially viable being close to Gladstone, and feasibility studies were undertaken by companies. The Stewart oil shale project has a history of changes to methods of retorting and serious environmental issues which will be further discussed in Chap. 7.

5.2 Mount Coolon Oil Shales

Abstract
The Mount Coolon oil shales of early Miocene age were deposited on a Devonian surface of uneven topography. They are a lacustrine deposit with cyclic deposition of organic-rich shales, lignites and kaolin clay. The organic matter contains mainly algal remains, some pollen of land plants, fungal spores and bacterial remains. Massive pyrite beds were encountered in the sequence at certain levels pointing to extreme anoxic conditions prevailing for a while. A pure white kaolin clay about 50 m in thickness overlies the organic-rich shales in all boreholes examined. This oil shale provides an interesting study of a segment within the climate of the time.

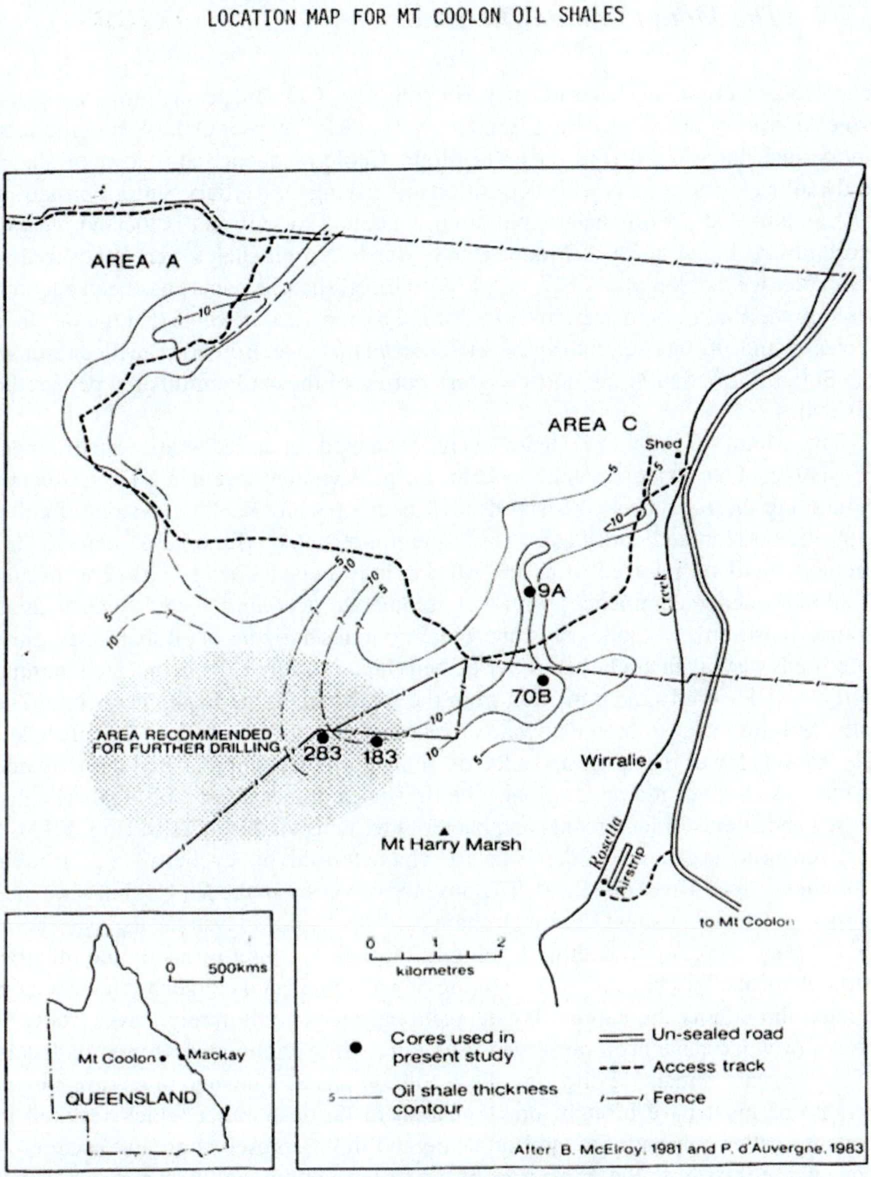

Fig. 5.4 Key map: Source Glikson (1982) Unpublished Report to I.M.C. (International Mining Corporation)

5.2.1 The Mount Coolon Oil Shales

The Mount Coolon oil shales of early Miocene age (J. Owen, pers.comm.) are lacustrine deposits of unknown extent, 250 Km west of McKay near Olderfleet homestead and Mount Harry Marsh (Fig. 5.4). The Mount Coolon sequence consists of oil shales and kaolin claystone in a cyclic deposition underlying the Tertiary Suttor Formation. The area drilled for oil shale exploration is about 126 km^2 and is located west of Rosetta Creek and north of Mount Harry Marsh. No oil shales were recovered in boreholes west of longitude 147 ° 12.5′. A thick oil shale sequence has been encountered to the south, south-west and west of the barren area. It seems that the oil shale facies extends further south to an as yet unknown distance. Boreholes drilled through the Sutter Formation in the north western corner of the exploration area penetrated oil shales.

The Mount Coolon oil shales were deposited in a lacustrine environment consisting of an extensive lake system, on a Devonian basement. The sequence containing the oil shales consists of thick beds of white kaolin claystone, lignite, a pyritic horizon and dark kaolin claystone impregnated with organic matter. This sequence will be referred to as the Mt Coolon Sequence (Fig. 5.5). The Mount Coolon sequence commences with a conglomerate overlain by sandstone, with an abrupt transition into kaolin claystone which contains horizons of oil shale and lignite inter-beds varying in thickness. A pyritic bed (Fig. 5.6) about 20 cm thick, containing pyrite nodules marks the transition from the sandstone to the kaolin claystone. The latter becomes grey to brown upwards to about 1 m below the base of the oil shales. The kaolin clay as it gets closer to the oil shale horizon is impregnated with organic matter which gives it a dark colour. The oil shales proper consist of algal remains, pollen and spores of land plants and bacterial remains visible in TEM (Fig. 5.11).

Lacustrine organic-rich deposits are characterised by cyclicity (e.g. Rundle Complex, Green River oil shale). The absence of varves in the Mt Coolon sequence points to lack of distinct seasonal changes. The kaolin claystone deposits point to a "temperate climate with high rainfall, through leaching of alkali and alkaline earth elements." (Levinson 1974). Mixing of water and good oxygenation must have existed throughout the kaolin clay depositional phase; only rarely leaves, roots or pieces of wood have been preserved. The pure white kaolin clay section is devoid of any organic remains. Following the Kaolinite phase, a change in environmental conditions must have brought on stagnation in the lake water which resulted in organic matter concentration and oil shale and lignite preservation and accumulation. Algae thrived in the upper part of the water column and their remains which fell into the anoxic bottom waters have been preserved. The algae consist mainly of *Botryococcus* remains that strongly fluoresce in UV excitation (Fig. 5.7; 5.15) while filamentous microbial remains are rare (Fig. 5.11) while fungal spores and fungal resting bodies are part of the organic input (Figs. 5.8; 5.9).

Denudation of the uplifted landscape resulting in low relief by the end of the weathering cycle with no climatic change brought about meromictic conditions in the lakes. Favourable condition must have prevailed to sustain a rich algal community

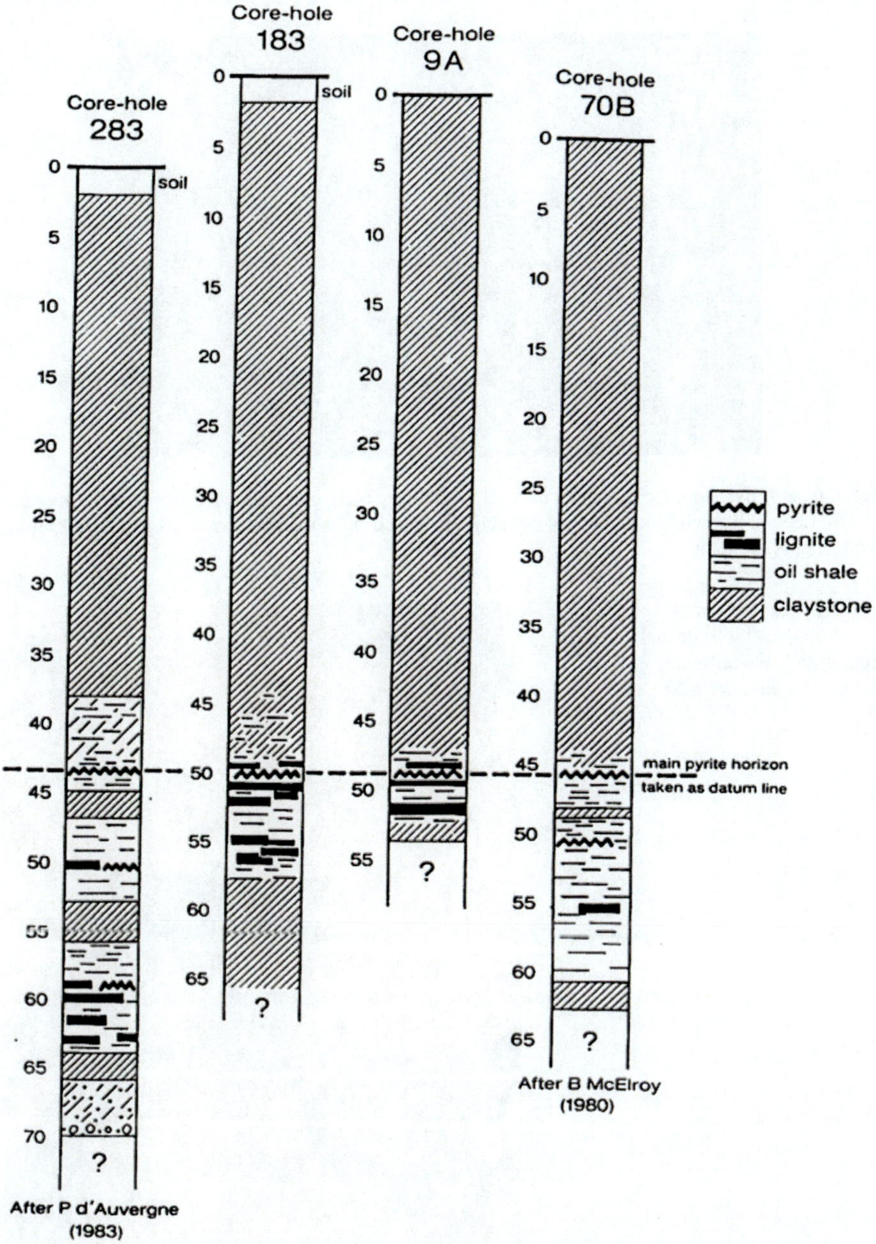

Fig. 5.5 Mount Coolon sequence: logs of 4 boreholes. *Source* Glikson (1982) Unpublished Report to I.M.C. (International Mining Corporation)

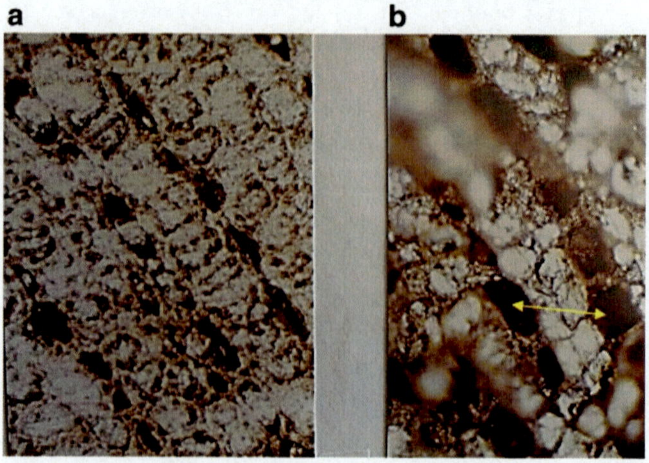

Fig. 5.6 a–b Pyrite replacing plant cell walls and filling plant cell cavities of wood tissue, remains of OM seen as brown-black patches (arrow). Reflected light microscopy. a taken at X200 mag. **b** taken at X250 mag

Fig. 5.7 Algal remains (*Botryococcus*) in reflected light with UV excitation; fluorescence mode. X200

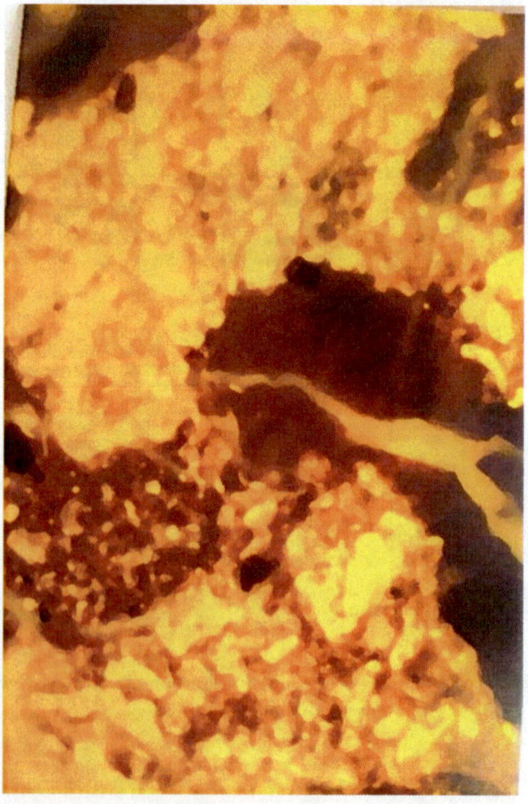

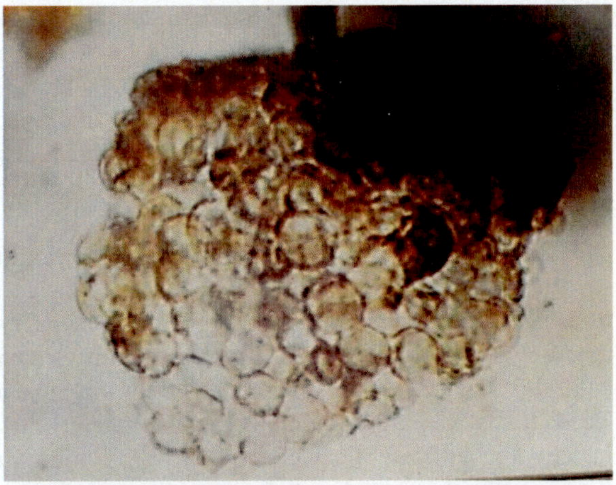

Fig. 5.8 Aggregate of fungal spores. Transmitted light microscopy. X400

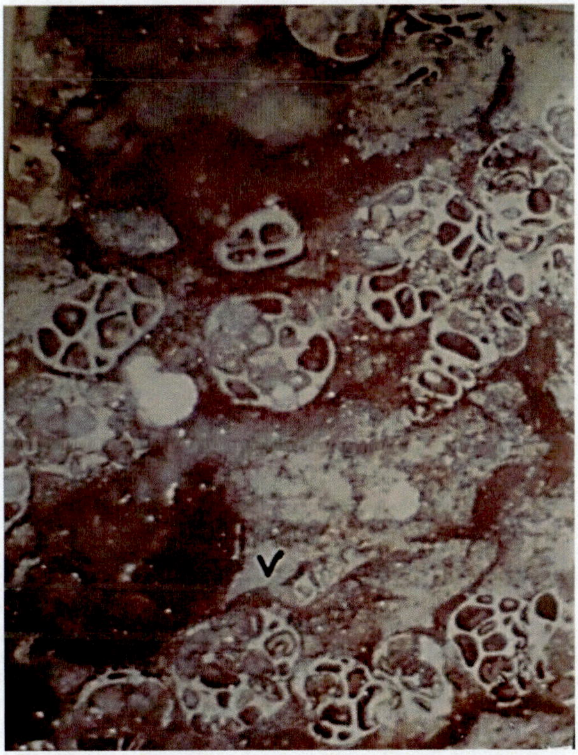

Fig. 5.9 Fungal resting spores. Reflected light. X200mag

in the upper part of the water column. Identifiable algae are *Botryococcus*, Tasmanite and filamentous cyanobacteria; details of most algal remains are not identifiable in light microscopy, but are clearly identifiable in TEM (Fig. 5.16a, b). Pollen of trees, and spores of ferns (Fig. 5.12) were deposited in the lakes and preserved in the anoxic sediments. Anoxic conditions in the lake prevailed for a time when structural or other factors brought on water turn, oxygenation and kaolin clay deposition with very little organic matter.

Kaolin clay and kaolinite form from weathering of aluminium silicate minerals such as feldspar, the solubility of which occurs in warm inter-tropical climatic conditions. Kaolin clay precipitates when pH rises since its solubility decreases rapidly as neutral pH values are reached. Humic acid particles with coccoid microbial remains and other organic matter remains make up the bulk of some samples; observed in TEM. It has been experimentally demonstrated (Playter et al. 2017) that clay sinks faster when adhered to bacterial cells. TEM Shows humic acids forming a network (Fig. 5.10) with coccoid microbes visible within this network. The humic acids may be playing the same role as that of clay particles, namely as carriers of microbes.

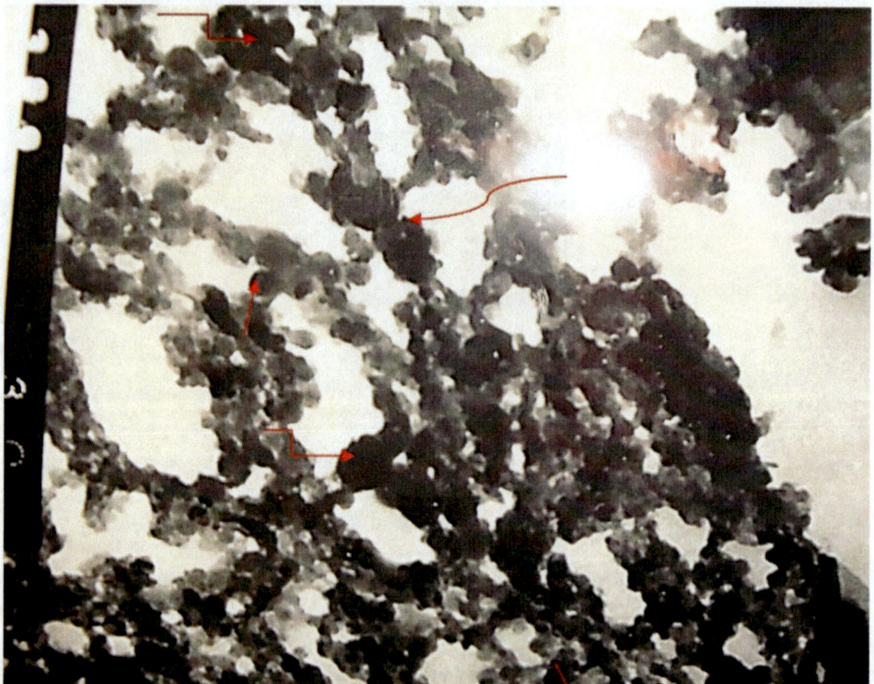

Fig. 5.10 Humic acids with occasional coccoid microbes (arrows) form a mesh work. Taken at X 30 000

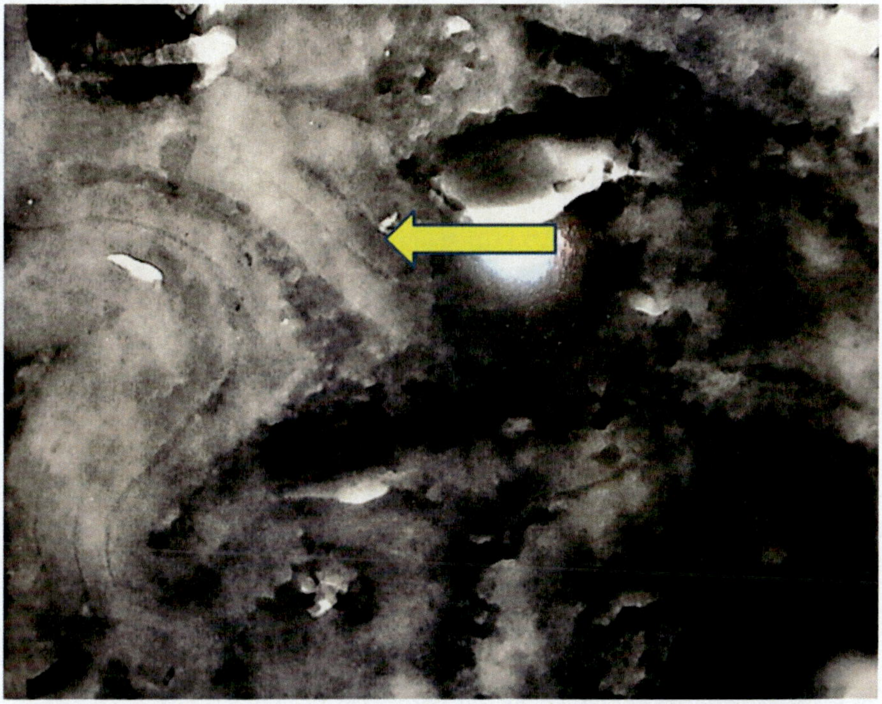

Fig. 5.11 Filamentous bacteria; X 108 000

Vacant spaces observed in TEM represent minerals dissolved during preparation and organic matter concentration (Fig. 5.10).

Pyritic horizons within the oil shale sequence mark return to anoxic conditions after an influx of sulphate rich marine incursion following the kaolinite deposition phase. The high organic carbon content of the organic rich sediments coupled with influx of sulphate from incursion of sea water enabled high activity of sulphate reducing bacteria, leading to formation of pyrite which often replaced plant cell walls amd filled cell cavities (Fig. 5.6).

Remnants of organic matter trapped in plant cells with pyrite filling most of the space (Fig. 5.6) support pyrite formation through bacterial activity. Anoxic conditions must Have prevailed during the pyrite formation and its preservation. An important factor in enabling thick accumulation of organic matter was a meromictic nature of the Mount Coolon lake system at times, namely no overturning of lake of water during the oil shale depositional phase. A eutrophic lake system supported a flourishing plankton population. Occasional lowering of the water table and partial reverting to swamp conditions at tmes is marked by significant input from fern spores (Fig. 5.12).

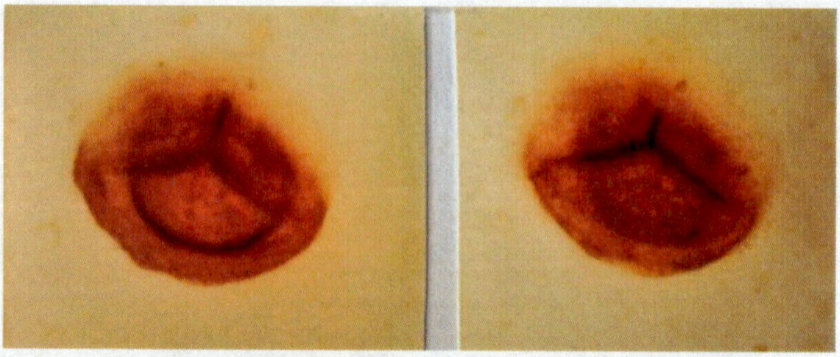

Proximal view of spore.

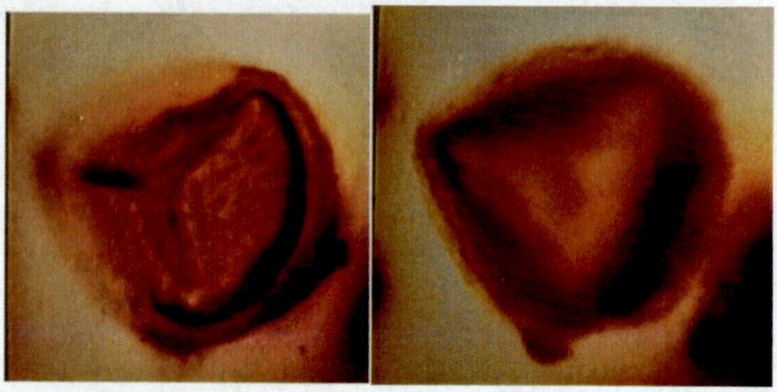

Proximal and distal view of spore

Fig. 5.12 Fern spores. Transmitted light, X400 mag

5.2.2 The Mount Coolon Sequence

A total thickness of the sequence cannot be determined from existing data as boreholes have not reached basement. Borehole 283 revealed at its base a clayey sandstone unit about 15 m thick, followed by a10m pure oil shale bed (Fig. 5.5). The oil shale is brown in colour and has high water content. When air dried, the oil shale acquires very low density, reminiscent of torbanites. The oil shale beds are overlain by pure white kaolin claystone about 3 m thick, with the exception of the contact area with the oil shale where presence of organic matter gives the clay a greyish-brown colour. Another oil shale bed, about 6 m thick follows, and is succeeded by another 2 m thick kaolin clay bed, which in turn is overlain by a 10 m thick bed of brown oil shale, terminated by a massive pyrite bed. The pyritic horizons can be traced throughout all boreholes, marking a depositional-environmental **event** throughout the basin. They have been used by the present author as a marker horizon, allowing correlation between the boreholes (Fig. 5.5).

Certain horizons are predominantly composed of fungal remains (samples 21 and 5) with very low Ro vitrinite OM (Fig. 5.13) which is made up of degraded

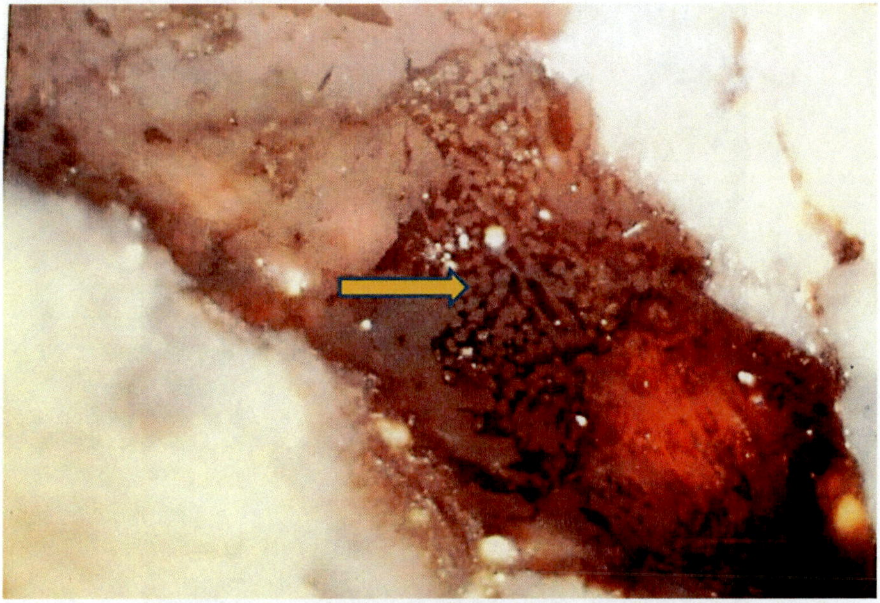

Fig. 5.13 Reflected light microscopy. Vitrinite of low reflectance ('Vitrinite-like'/ Vitrinite B) (Ro = < 0.4%. Bright spots are pyrite and white areas are calcite. Notice vitrinite breaking up into proto-humic acids (arrow) X400

woody plant remains. Botryococcus colonies are disperesed within the low Ro vitrinite (Fig. 5.14). Pollen of various species of the genera *Dacridium*, *Nothofagus*, and *Casuarinidites* (J. Owen, pers.comm.) represent allochthonous input from land plants.

The brown oil shale sequence of lacustrine origin is dominated by algal remains which can be seen as strongly fluorescing bodies in reflected light microscopy (Fig. 5.14) and with UV light excitation (Fig. 5.15). TEM however enables more accurate identification, namely that of the alga *Botryococcus;* the best -preserved parts of this alga visible are its lipid bodies (Fig. 5.16a, b).

While *Botryococcus* is a main component of the oil shales proper, the lignite beds contain mainly pollen exines and low Ro vitrinite. The pyritic horizons contain humic acids and possible filamentous bacteria (Fig. 5.11).

The carbon isotope values of the mount Coolon oil shale deposit can be divided into three main groups: **IA**. The coaly lignites with values of −26.57‰ and 26.86‰.

I. IB. Lignite samples having values of −27.58‰/−27.24‰, admixed with some algal/oil shale material.
II. II−28‰ and −29.7‰ values for mixed samples of lignitic/coaly material and higher input of algal oil shales than group IA.
III. III. A third group of pure algal oil shalcs, having the highly ^{13}C–depleted values of −33.9 to −35.9‰.

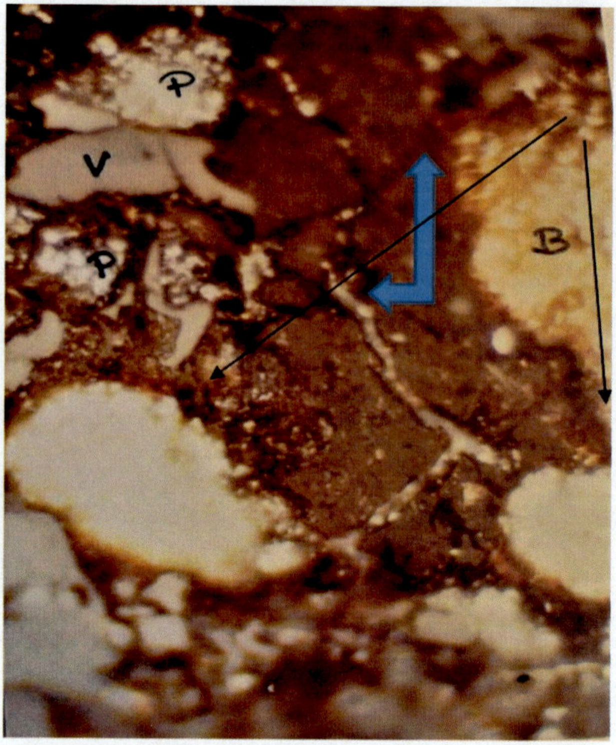

Fig. 5.14 Reflected light: low Ro Vitrinite Ro = 0.2% (blue arrow), high Ro Vitrinite (V), *Botryococcus* colonies (B, black arrows), Pyrite (P)

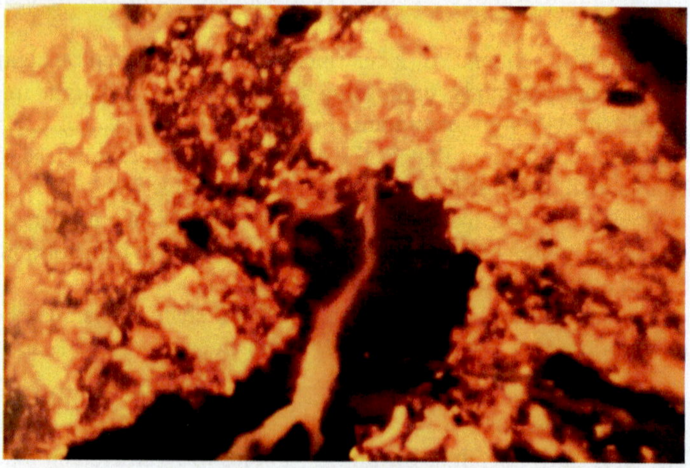

Fig. 5.15 Algal colonies (bright yellow), brown material is past oil generation. Reflected light, UV excitation. X200

a

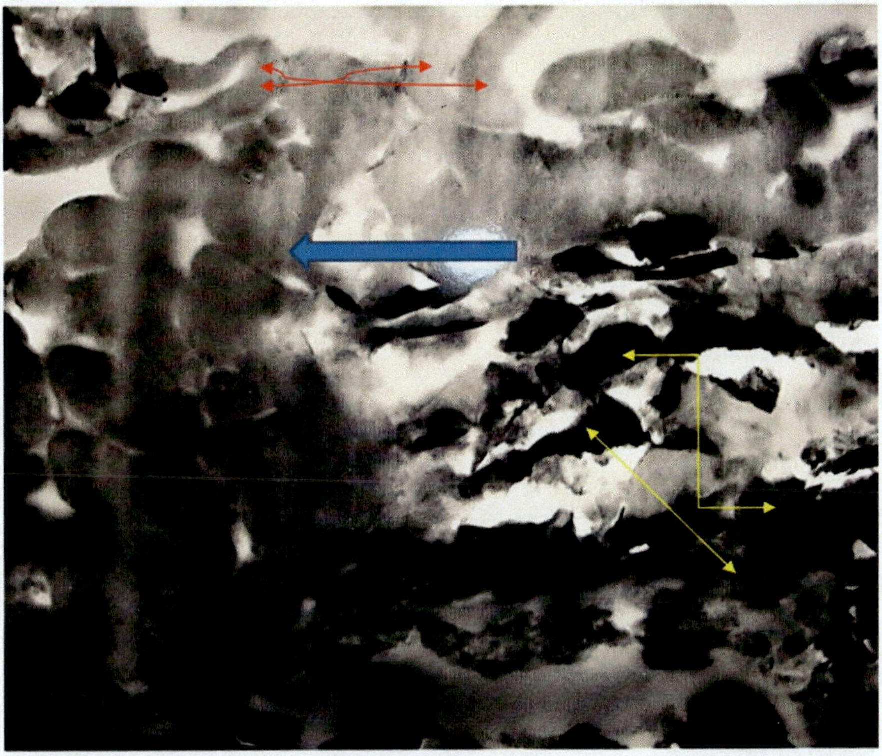

b

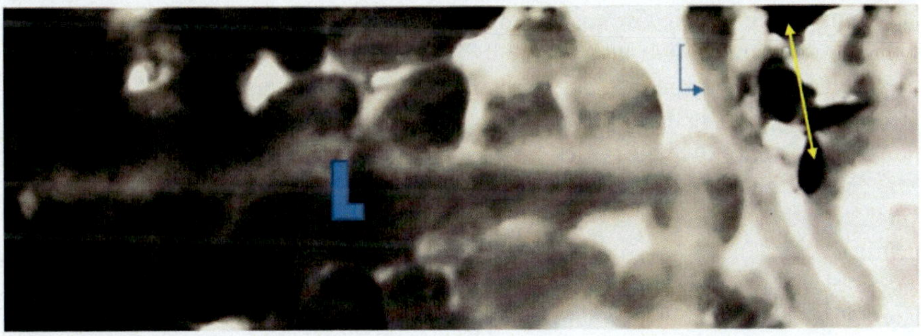

Fig. 5.16 a–b: TEM showing lipid bodies of *Botryococcus* (L) and exines of pollen (thin blue arrow), X45 000. Electron opaque particles (thin yellow arrow) are metals absorbed on to OM

Table 5.3 C–isotope composition of Mt. Coolon OM

Sample	$\delta^{13}C‰$
9A/2–Brown algal oil shale	−35.94
9A/5–Clayey oil shale	−33.91
283/5–Brown algal (43.6 m)	−29.71
283/6–Brown and black shale and lignite	−28.26
283/10–Brown algal (57.7–58.1 m)	−28.23
283/11–Brown clayey, waxy/algal (59.2–59.6 m)	−28.46
283/13–Brown waxy algal (60.2–60.7 m)	−28.26
283/14–Dark brown-black coaly (60.7–61.1 m)	−27.24
283/16–Brown clayey-waxy (62.1–62.6 m)	−28.83
283/19–Black lignite (64.5 m)	−26.86
70/4–Brown oil shale with coaly particles	−28.23
70/5–Brown oil shale " " "	−28.33
70/8–Brown oil shale.Algal (52 m)	−28.33
70/9–Brown lignitic oil shale " " "	−27.58
183/1–Brown, algal oil shale (53.1–53.5 m)	−28.58
183/2–Lignite (54.9–55.3 m)	−26.57

The three isotope main groups (Table 5.3) are an expression of the predominant organic matter type and its diagenic history in each set of isotope value groups. Group I (A&B) represent input of land plants in the lignitic parts of the deposit. The differences between the sub-type A & B being in types of vegetation; spores of ferns and fungi or tree pollen. Groups II and III emphasise input of algal components.

It is evident from the major elements analysis (Table 5.4) that carbon varies in direct relation to the organic matter content. The highest C content is in samples from Drill-hole 70A, in algal –dominated oil shales with input of some lignitic material. Samples in proximity to the pyritic horizon yielded high S, while most other samples were relatively low in S. Carbon and S seem mostly positively correlated (Fig. 5.17) which would be expected as sulphate reducing bacteria were active where OM was plentiful.

Samples with high ash and low Carbon are unsuitable for H/C determination, as they are by definition not a hydrocarbon source. However, the samples with high C and H and low ash were generally in the 'Kerogen' type II regime on the Van Krevelen diagram, as is common in OM with significant terrigenous input.

Carbon and Sulphur are mostly positively correlated as expected where the organic matter played an important role in sulphate reduction and eventually pyrite formation. There is an anomaly in a sample with 40% carbon and highest sulphur composition where an inverse relationship is apparent. One other high C sample with very low S shows a similar inverse trend. In these samples the S is tied up in pyrite.

Table 5.4 Major elements

Major elements							
Sample	C	H	O	N	S	Ash	H/C
D.H.70B (45–48 m)							
COO-1	47.72	5.65	16.97	0.36	0.58	18.34	n.a
COO-2	58.04	5.94	21.31	0.85	2.85	8.28	1.22
COO-4	59.21	6.40	20.87	0.49	0.92	5.69	1.31
COO-6	52.80	5.48	17.70	0.70	1.11	9.34	1.24
COO-7	41.15	4.57	17.24	0.20	**7.94**	25.21	n.a
COO-10	66.11	7.11	16.64	0.18	2.40	2.18	1.28
COO-11	51.61	5.72	20.34	0.36	1.28	11.69	1.32
D.H.9A (50–53 m)							
COO9A-2	35.77	4.57	16.26	0.38	1.35	39.31	n.a
COO9A-3	20.67	2.91	18.20	0.23	2.93	51.39	n.a
COO9A-5	39.60	3.89	24.02	0.43	2.05	24.22	n.a
D.H.183 (50–57 m)							
183/1	46.66	4.77	23.01	–	**5.03**	15.06	1.22
183/2	22.63	3.54	12.52	–	0.82	50.07	n.a
183/3	34.19	5.16	13.52	–	0.74	38.15	n.a
183/20	43.03	4.58	25.05	–	**5.84**	15.99	1.27
183/21	47.13	5.95	15.82	–	1.21	23.58	
D.H. 283 (selected samples with high OM)							
283/12	55.41	5.76	22.19	–	1.81	9.63	1.24
283/14	46.49	5.55	23.94		1.10	11.92	1.42
283/19	43.66	4.78	23.78	1.64	13.90	1.30	

Trace Elements in Mount Coolon Samples

7 samples were analysed for trace elements. The samples were picked from oil shales mixed with lignite. High Ti was obtained from all samples, and ranged from 604 to 4015 ppm. Vanadium was relatively low ranging from 9 to 300 ppm with the average of 130 ppm. Cr was low ranging from 10 to 69 ppm.

5.3 Summary

The Miocene Mount Coolon oil shales, the youngest known oil shale deposit in Australia, show unique characteristics in terms of variability of organic components, elemental and isotopic compositions. It is composed of a cyclic sequence containing thick claystone beds alternating with lacustrine/swamp derived organic-

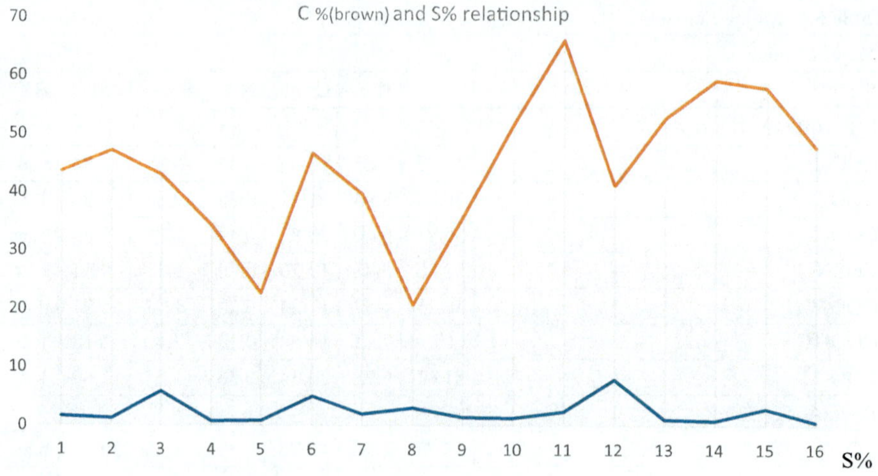

Fig. 5.17 Relation between carbon and sulphur in mount Coolon oil shales

rich beds. The claystone is a pure kaolinite, whereas the organic matter in these oil shales is a mixture of algal remains, fungal remains and pollen of swamp/land vegetation. The contribution of each component was dependant on environmental conditions within the source area, whether lakes and high- water levels, onset of drier conditions and reverting to swamps with their fluctuations in water levels and resulting vegetation. The swamp deposits are characterised by the abundance of wood/cuticles and fern spores (Fig. 5.12) which would have come from an in situ source of the swamp vegetation. Interestingly these cyclic facies are also expressed by their isotopic compositions,

The cyclic deposits of the Mount Coolon oil shales point to tropical conditions with rejuvenation of the landscape through tectonic activity at times. This was followed by intense weathering in the highlands and deposition of the kaolinite in the basin in a tropical climate.

Kaolinite deposition was phased out with advanced erosion of the landscape and moderation of gradients, followed by meromictic conditions in the lake system. Planktonic algae and organic debris from surrounding swamps sank into an anoxic environment where they were preserved. Pyritic horizons can be traced in the bore-holes throughout the basin. Pyrite in lignite beds has replaced cell walls of wood and filled cell lumens. These horizons mark short term marine incursions into the basin, probably brought on by structural movements. The pyritic horizons are followed by an onset of kaolin clay deposition, another cycle in the sequence formation.

References

Baker, W. E. (1973). The role of humic acids from Tasmanian podzolic soils in mineral degradation and metal mobilization. *Geochimica Et Cosmochimica Acta, 37,* 269–281.

Cheshire, M. V., Barrow, M. L., Goodman, B. A., & Mundie, C. M. (1977). Metal distribution and nature of some Cu, Mn and V complexes in humic and fulvic acid fractions of soil organic matter. *Geochimica et Cosmochimica Acta, 41,* 1131–1138.

Coshell, L. (1983). Cyclin depositional sequences in the Rundle oil shale deposit. Processes 1st Austral. Workshop on Oil Shales: 25–30. Lukas Height.

Dale, L. S., & Fardy, J. J. (1983). Trace element characterisation in oil shales. Processes 1st Austral. Workshop on Oil Shales: 57–60. Lukas Heights.

Glikson, M. (1982). Report to I.M.C. (International Mining Corporation), unpublished.

Henstridge, D. A., & Hutton, A. C. (1987). Geology and organic petrography of Nagoorin oil shale deposit. *Fuel, 66,* 301–304.

Levinson, A. A. (1974). *Introduction to Exploration Geochemistry.* Calgary: Applied Publishing LTD.

Moore, A. J., Wright, B. C., Redann, P. J., & Gannon, A. J. (1987). Implications from several oil shale projects sdudies. *Fuel, 66,* 298–300.

Noon T. (1987). Oil shale exploration activity in Queensland during 1985/86. *Fuel, 66*(3), 295–297.

Patterson, J. H., Dale, L. S., Fardy, J. J., & Anthony, R. (1987). Charaterization of trace elements in Rundle and Condor oil shales. *Fuel, 66,* 319–322.

Playter, T., Kornhauser, K., Owttrim, G., Hodgson, C., Warchola, T., Mloszawska, A. M., et al. (2017). Microbe-clay interactions as a mechanism for the preservation of organic matter and trace metal biosignatures in black shales. *Chem. Geology, 459,* 75–90.

Pope, G. J. (2000). The application of sequence stratigraphy in modelling oil yield in the stuart oil shale deposit. M.Sc. Thesis, QUT, published. p. 146.

Veevers, J. J., Powell, C., & Roots, S. R. (1991). Review of Seafloor Spreading around Australia. *Australian Journal of Earth Science, 38,* 373–389.

Chapter 6
Discussion and Conclusions in View of the Methodologies Applied in This Study of Oil Shales

Abstract This chapter highlights the combined techniques and methods used throughout this book to understand oil shales; the organic and inorganic components within these deposits and the relationships between them. Thus, it is demonstrated that not one unique analytical method is sufficient to understand the geological evolution and properties of oil shales, but a combination of an array of analytical and observational methods on different levels enables their understanding, economical evaluation in terms of prediction of the oil yield, and expected environmental problems following oil extraction.

The TEM observations of the organic matter elucidated its composition when optical microscopy was beyond resolving power. Differentiation between algal remains and degradation products is possible using metal stains. Algal remains do not absorb metal stains (e.g. lead citrate) as has been shown in the initial experimental stage of the use of TEM to observe differences between organic entities.

Various kind of bacteria are distinguishable in TEM; these include filamentous microorganisms and coccoid forms. Humic acids (HA) cannot be recognised in optical microscopy, but are a clearly distinct entity separate from other organic matter in TEM. Examination of this alkali-soluble fraction revealed an ultra-structure consisting of minute globules forming aggregates (Fig. 2.5 in Chapter 2). Stages of humic acid formation may often be followed in TEM, from breakdown of filaments into smaller and smaller particles.

Allochthonous or autochthonous source of the humic acids is consistent with their mode of distribution within the rest of the organic matrix. It was observed that when HA were irregularly distributed throughout the rest of the organic matter as isolated globules or small aggregates, an insitu origin was supported by carbon-isotope composition. However, when HA were observed in TEM to form concentrated accumulation within spaces between the rest of the organic matrix, carbon isotope composition points to an allochthonous source.

TEM observations of torbanites showed that the non-fluorescing parts of the OM as well as parts displaying brown fluorescence are made of degraded algal remains, bacterial remains and humic acids. The non-degraded *Botryococcus* colonies in TEM were typically homogenous in texture, electron transparent with often lipid bodies

M. Glikson-Simpson, *Oil Shales*, https://doi.org/10.1007/978-3-030-60675-6_6

recognisable as separate entities. The cell walls of this alga could be traced by its aligned electron opaque minute grains characteristic of this alga (TLS of Largeau). Recent rubbery accumulations of *Botryococcus* in Coorongite displayed a very close resemblance to the algal remains in torbanites.

The structures of OM of the Rundle-Condor oil shales were clearly identified in TEM as that of the green colonial alga *Pediastrum*. In these oil shales, electron opaque areas associated with humic acids proved to be mostly Titanium. The HA fraction proved to be allochthonous of a terrestrial/swamp vegetation source; Ti sourced from weathered volcanic rocks washed into the lakes. The Rundle complex oil shales of early Eocene age would have been deposited in a vast lake system, under humid conditions as inferred from high Ti/Al ratio.

The much-debated nature of the Green River oil shales became clear in TEM observations, as the OM displayed the characteristics of non-degraded algal remains with pockets of HA concentrations associated with metals. The association of Cr, Ti and Fe has been shown to be dominant in the humic acids fraction of this oil shale. Carbon isotope composition of the OM and HA supports an allochthonous source for some of the humates. Other electron opaque particles of metals and pyrite were dispersed throughout the predominantly algal OM. The mineral particles display flow-lines which give an impression of movement while in a colloidal or other semi-liquid state.

It is clear from this study that biodegraded OM and cyanobacterial remains tend to produce lower H/C ratios than non-degraded *Botryococcus* or *Pediastrum* accumulations. The oil shales with highest oil yield are therefore those with green algae as their predominant organic component. This is demonstrated by the difference in H/C ratios of algal dominated Rundle oil shales (1.5) compared to lignite dominated Condor (0.9), both are part of the Rundle Complex of central QLD coast Eocene oil shale deposits.

The wide variation in respect to type of OM and mineral composition is interpreted as a function of the climatic conditions prevalent in the depositional environment at the time. On Cretaceous paleogeographic maps Australia was rotated 90° eastward from the present location and was situated between latitudes 80° S and about 50° S. The Eromanga Basin was open to the ocean in the east, with a warm current flowing along the eastern margins of Australia at the time. The Cretaceous period experienced higher temperatures than at present over a wide range of latitudes. It is believed that tropical to sub-tropical conditions extended to 70° Latitude, with an arid zone at 55° Latitude. As noted earlier, the temperature of the surface and near surface water during the Toolebuc depositional phase had to be in the range of 18–24 °C. The postulated salinities were 35 ppt and the pH around 8. These environmental conditions were optimal for the growth of coccoliths whose remains are present in abnormally high concentrations throughout the Toolebuc oil shale facies. No direct evidence is available regarding the temperatures and salinities of the bottom waters. Only indirect assumptions can be made from the existence of cyanobacterial mats. These point to relatively warm waters such as exist today in Spencer Gulf (South Australia), Shark Bay (Western Australia), Laguna Guerrero Negro (Mexico) and Solar Lake (Red Sea).

The precipitation of carbonate within the cyanobacterial mats indicate high pH (above 7). Salinities associated with such mats at present are higher than those of average water, e.g. hypersaline in Solar lake (Krumbein et al. 1977), 50 ppt in Shark Bay and 40–45 ppt in Spencer Gulf (Burne et al. 1980). The water temperature reported for Spencer Gulf is between 12 and 25 °C and for Shark Bay 18 and 28 °C. The temperature assumed for the surface waters of the Toolebuc sea, between 18 and 24 °C are in the range of today's habitat of cyanobacterial mats. Assuming salinities of about 35 ppt for surface waters of the Toolebuc sea, the denser bottom waters would be expected to have higher salinities perhaps 40–50 ppt. Undisturbed shells abundant in the Toolebuc Formation indicate a quiet water environment. The rich plankton in the upper water column brought about depletion of oxygen in lower parts of the water.

High Nitrogen in the organic matter of the Toolebuc oil shales may further support cyanobacterial activity. The diagenetic carbonate has been deposited as a result of micro-environmental changes in the cyanobacterial mat, hence the close association between the low Ro vitrinite and carbonate. Algal-dominated oil shales such as torbanites, Rundle-Condor and Mount Coolon do not contain carbonates, have consistently low nitrogen and do not support cyanobacterial mats. Cyanobacterial mats in the modern marine environments are major contributors to OM in sediments, due to their being impermeable and creators of anoxic conditions. On the other hand, lacustrine environments create conditions favourable for algal blooms of *Botryococcus* and *Pediastrum*, both highly resistant to bio-degradation, which makes them major precursors of richest oil shales.

During the Cambro-Ordovician times, Australia due to its N-S extension extended from the equatorial region to 30° latitude. Open marine shelf conditions are suggested here for the Middle Cambrian organic-rich deposits, thus placing the Camooweal oil shales in the anoxic silled-basins on oceanic shelves according to the criteria of Damaison and Moore (1980). The two marine oil shales; namely the Toolebuc OM and the Camooweal show close resemblance. In both deposits, carbonates are interbedded with OM and calcite, sharing a similar diagenetic origin.

It is evident from this study and its conclusions that the nature and origin of the 'amorphous' OM in oil shales is best understood by combining the results of several methods and techniques. Thus, TEM allows the recognition of organic components unresolvable by optical microscopy. Carbon isotope analysis demonstrates consistent relationships between the OM components of the oil shales and the source organism, in spite of diagenetic processes, which altered, and in some instances erased the original compositions.

The combination of TEM and carbon isotope analysis enables the identification of the specific organic fractions and their association with specific trace elements as well as giving an indication as to the source of these metals.

Comparison of the oil shale deposits described in this book with analogues in present day depositional environments with high primary productivity can add to the understanding of past deposits. Various fresh water bodies containing *Botryococcus* blooms have been compared to torbanite deposits.

Spencer Gulf, the Gulf of Mexico and the microbial mats of Abu Dhabi may in some way be analogous to the Toolebuc oil shales and Camooweal oil shale deposits.

The central African Great Lakes system is a possible analogue of the Rundle-Condor-Stuart organic-rich deposits. Studies of the OM in these environments, using the same techniques and methods as applied to the oil shales would enable direct comparison.

References

Burne, R. V., Colwell, J. B., Pain, L., & Tratt, M. M. (1980). Sedimentological studies in the Spencer Gulf. *Baas-Becking Laboratory Annual Report, 1980,* 6–11.

Demaison, G. J., & Moore, G. T. (1980). Anoxic environments and oil source bed genesis. *AAPG,* 1179–1209.

Krumbein, W., Cohen, Y., & Shilo, M. (1977). Solar Lake (Sinai) stromatolitic cyanobacterial mats. *Limnology and Oceanography, 22,* 635–656.

Chapter 7
Mining and Processing Oil Shales for Energy and Petroleum Production

Abstract Oil shales provide through their cyclicity excellent scientific knowledge of past climate and environmental conditions throughout geological times. However, they do not present a convincing rationale as a mined resource for energy and oil production. The reasons will be outlined in the following chapter.

7.1 Environmental and Health Effects of Oil Shale Mining and Processing

Oil shale mining presents possibly one of the greatest environmental problems of any resource, due to the extent of the deposit to be mined to obtain a few drops of oil, and the mode of mining. Mining above ground of oil shales requires huge quantities of rock to be excavated and extracted to produce the oil. The quantities vary slightly from one deposit to another, but generally are immense. As seen in the Australian examples of potential oil shale resource such as the Rundle complex where a ton of rock produces at the most 82 L of oil/ton. In Duaringa Basin a seam produced 105 L/ton. Generally, the ratio of Waste: Ore has been reported to be 6.9:1 (Dixon and Pope 1987) which can be rounded to 7:1. In other words, to obtain 1 L of oil 7 tons of rock have to be mined. Some precious metals produce less than a gram from a ton of rock, but the price (e.g. gold) makes it economical. In the case of oil shales, it is far from being economical. All oil shales production is heavily subsidised by Governments in Estonia and the US and other countries, and was financed by shareholders in the Australian pilot project. Furthermore, the 'productive' horizons in the Duaringa Basin for example (part of the Rundle complex) are 500 m deep. The cyclicity of these deposits means that only a fraction of the mined rock is oil yielding. The oil shales in a strict sense are thin beds alternating with thicker beds of lignite or clay, mudstones or other sediments lacking enough OM to provide a source of oil.

Schmidt (2003) assessed the Rundle-Stuart deposits (Fig. 7.1) and provided a detailed account of their potential in terms of oil production, as well as details of the demonstration plant and an optimistic view of its sustainability in regard to health and environmental impact, considering that the demonstration plant was only very small compared to a full-size mining and processing operation.

M. Glikson-Simpson, *Oil Shales*, https://doi.org/10.1007/978-3-030-60675-6_7

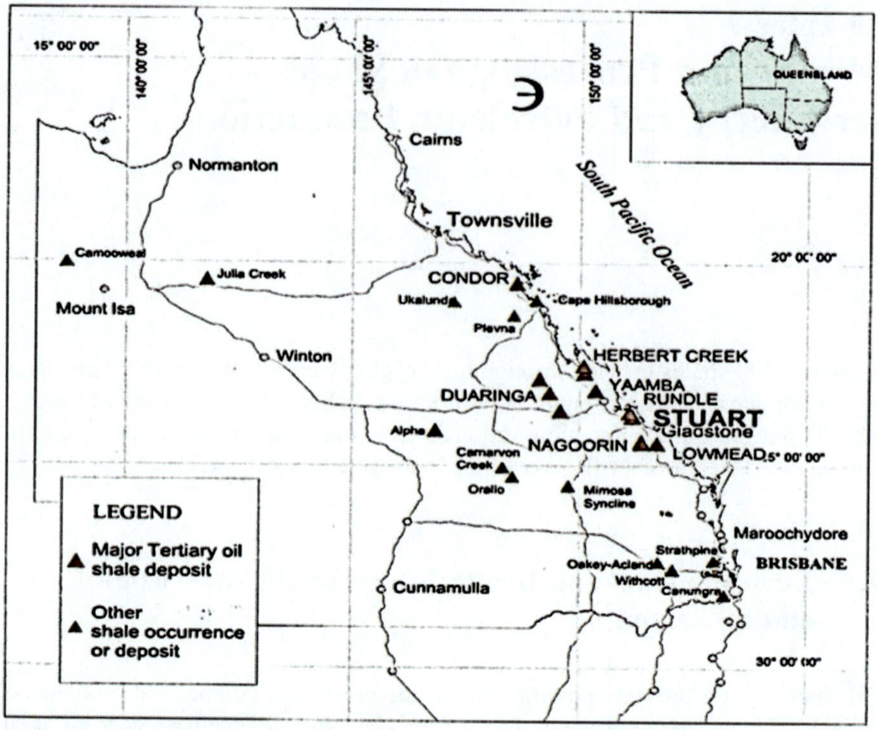

Fig. 7.1 Oil shales in Australia. *Source* Geological Survey of Queensland

Environmental impact studies were presented to the government at the time by Greenpeace (2000) as well as a public Submission report (D. Franks). The Queensland Government ordered an environmental impact study of the demonstration plant of the Stuart oil shale near Gladstone. The environmental review contains detailed studies of all aspects of these oil shales' mining, retorting, upgrading, spent shale and waste water management; leachates from the spent shale, hydrological problems, air quality and rehabilitation of the land. Following the detailed Report, the QLD Government placed a Moratorium on any oil shale developments in the area, until 2028.

It will be mentioned here that a rehabilitation exercise of spent shale from processing of the Stuart deposit showed that natural vegetation was established similar to the revegetation of coal mines in Bowen Basin before they follow the dip underground. However, the Stuart oil shale project was a small demonstration plant with a small area effected compared to a full-blown oil shale processing project that would pose a daunting task of rehabilitation.

Unlike coal which is an organic rock with primary mineral content between 0 and 4% (e.g. Permian Bowen Basin coals; Glikson et al. 2000). In terms of energy output; ***"Oil shale is vastly inferior to coal, firewood and even manure."*** (Grunewald 2006). Organic matter makes up between 10 and 20% of the productive beds. The rest is mineral matter. The high OM containing-beds are intercalated with thicker organic poor beds. In order to produce oil comparable to petroleum pumped from underground reservoirs, the shale goes into retorts, where essentially it is heated to temperatures between 500 and 850 °C depending on the type of retort. The produced hydrocarbons are not like crude oil pumped out of natural oil reservoirs but have to be upgraded by hydrogenation (adding hydrogen), to eliminate heterocyclic compounds and stabilize the oil before it is sent to a refinery. It is an energy demanding process which adds yet significant cost to the utilization of this resource.

In Estonia where the main energy sources come from oil shales the shales are not only processed for oil but are also being used directly as combustion fuel in power stations. Direct combustion of oil shales, apart from causing significantly higher CO_2 emissions than coal, causes corrosion and naturally leaves a residue of ash made up of fused minerals: 'Around 5 million tonnes of this residue are produced in one year in Estonia' (Koel 1999). Raukas and Punning (2009) reported radioactive elements in the ash from thermal combustion of oil shales in power plants deposited in NE Estonia. The oil shale dumps cover more than 20 km^2 and the radiation from radioactive isotopes in the oil shale dumps is many times higher than the background values in the soil (Raukas and Punning 2009).

Generally, the spent shale from oil shale retorts such as the Estonian ones is 30% greater in volume than the original mined rock. These are piled high up to over 60 m and termed "Estonian Alps" (Lotman 2016) and are visible from space. As there is some (2–5%) OM matter left in the spent shales they tend to spontaneously combust and cause fires. The spent shale dumps are leached and the harmful chemicals and trace elements pollute waterways and surrounding land.

Oil shales acquire a wide range of trace elements throughout their depositional history. As each oil shale deposit is unique in terms of its depositional environment, different trace elements will be concentrated in different deposits, as has been shown in the previous chapters. The mode of distribution of the metals in the different oil shales will have a bearing on their fate in the retort and subsequently on their mode of release into the environment. The mode of concentration of metals in the Australian Rundle Complex oil shales is predominantly in the humic acid fraction of the OM as observed in TEM coupled with EDS. Humic acids in these oil shales constitute about 30% of the OM in Rundle oil-shales and higher in Condor oil shales. The HA fraction with its adsorbed metals will be part of the spent shales after retorting. TEM studies of spent shales showed the metals intimately associated with fused humic acids (Glikson et al. 1987).

As can be seen in Table 7.1 Some of the toxic trace elements double and triple their concentration in the spent shale compared to the raw shale.

It is clearly evident from the above-mentioned table that the most harmful elements seem to become highly concentrated in the spent shales. The reason being, as mentioned earlier, is in their residence within humic acids, namely the non-oil yielding fraction of the OM. All elements with the exception of Sr show a significantly higher concentration in the spent shale compared to the raw shale. Needles to point out that all these elements are toxic and harmful to humans and other living things. The central Queensland oil shale venture was in a particularly sensitive area where a

Table 7.1 Trace element concentrations in Rundle; comparison between the raw and the spent shale

Sample	Elements ppm												%	
	Mn	Rb	Sr	Pb	Th	U	Zr	Nb	Y	Ni	Cu	Zn	TiO_2	Fe
R1	605	48	276	8	5	1	142	6	16	39	25	66	0.577	5.0
S1	1130	86	226	25	10	2	148	10	29	46	43	132	0.703	7.9
R2	825	51	389	8	5	1	106	6	17	41	26	60	0.575	1.4
S2	1160	85	225	25	10	2	151	9	28	43	48	119	0.714	8.1

R = Raw Rundle shale
S = Spent Rundle shale

narrow channel divides the resources from Curtis Island (Map Fig. 7.2) with its unique ecosystem, the mangroves fringing the coastline and serving as breeding grounds for prawns and fish. The latter are a source of livelihood for the local fishermen. Farms in the area have already been affected by the pilot plant. The heritage—listed Great Barrier Reef is north of the oil shale area and would be prone to pollutants from such a venture.

Oil from shale has been produced in 1926 under Japanese rule in Fushun. The oil production ceased after World War II and was resumed later. In 1961 China was producing one third of its total oil from shale. Production ceased after oil discovery, but resumed production in 2006 in a very large operation. Spent shale and retort shale ash are used in cement and brick production. No details are available on land reclamation or revegetation. In view of China's mega-dams it is unclear why would there be a need to use oil shales for power generation.

Jordan is another country that uses oil shales for electricity generation. Jordan does not have any natural energy resources but is blessed with a lot of sunshine and large empty spaces well suited for Solar Energy production. Electricity generation in Jordan is beginning to use gas, a significantly cleaner energy source, imported from Israel's off-shore wells. Brazil has been producing oil from shales for a long time.

The technique used to process the oil shales and its resulting environmental effects discussed above is the commonly used above ground mining and retorting or 'surface processing'. The other technique is in-situ retorting currently considered for the United States Green River **oil shales** (Fig. 7.4). This technique is similar to the one schematically illustrated in Fig. 7.3 for **shale-oil**/gas except of the additional heating process required for the 'immature' oil shales. There is no above ground mining involved but heating the oil shales and its upgrading is conducted underground by injecting steam or an oxidizing gas (air-enriched with oxygen) to generate heat. Major risk to ground water is from leaking oil and release of retort water that contains ammonia, phenols, lead, mercury and arsenic from the pyrolysis process. Ground water contamination is becoming more and more of a serious problem worldwide as natural supplies of drinking water are dwindling. In 2002, the oil shale fired power station industry in Estonia used 91% of the total water consumed by that country (Craig 2011).

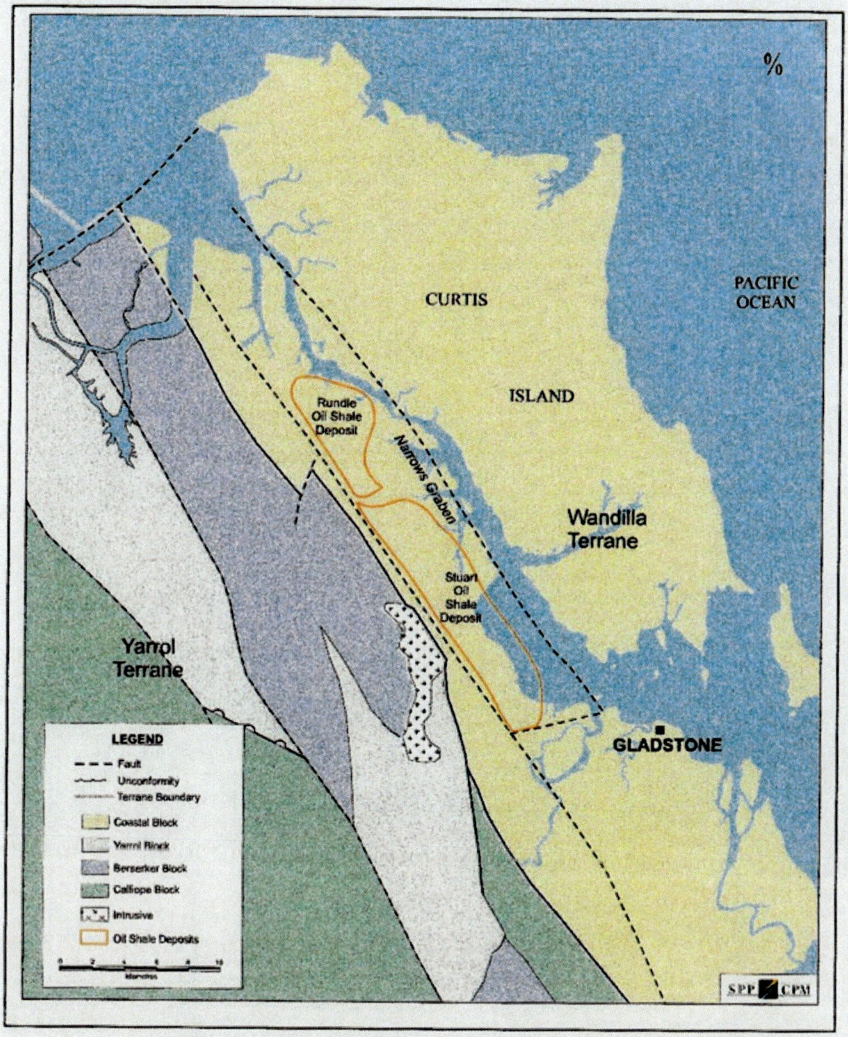

Fig. 7.2 Environment-sensitive areas around Rundle-Stuart oil shales prospects. *Source* Pope (2000)

It needs to be mentioned here that there is a difference between **shale-oil** and **oil-shale**. While oil shales have not yet generated oil naturally and therefore have to be heated at high temperature to create artificially oil generation (Fig. 7.4); shale oil has past the oil and gas generation stage naturally, but has retained it in cleats and fine fractures within the shale, similar to the retention of gas in coal seams. Shales are tight sediments unlike the porous sandstones or limestones. To release these hydrocarbons which are tightly attached to the rock, fracturing is used as in Fig. 7.3. Shale oil is being produced in significant amounts from Permian basins in

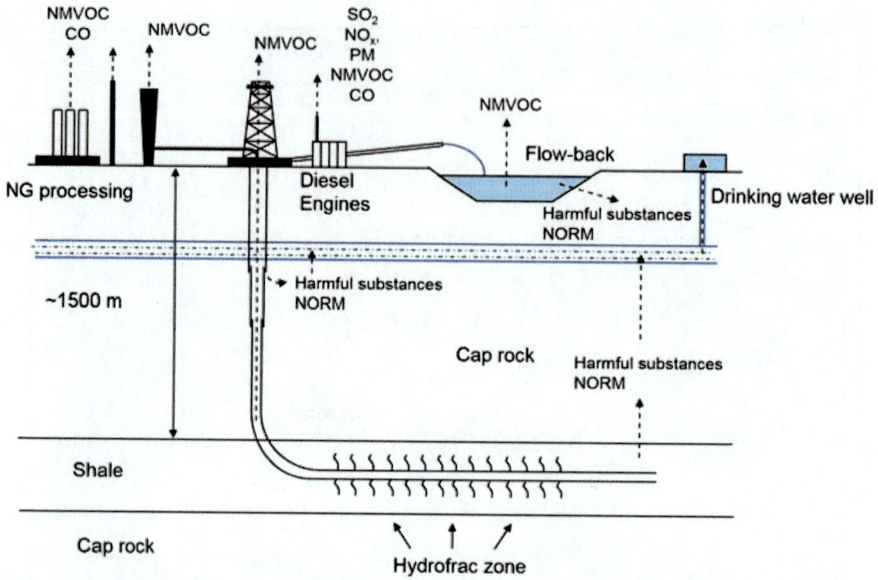

Fig. 7.3 Fracturing used in the insitu recovery of SHALE OIL/GAS (I/AENVI/ST/2011)

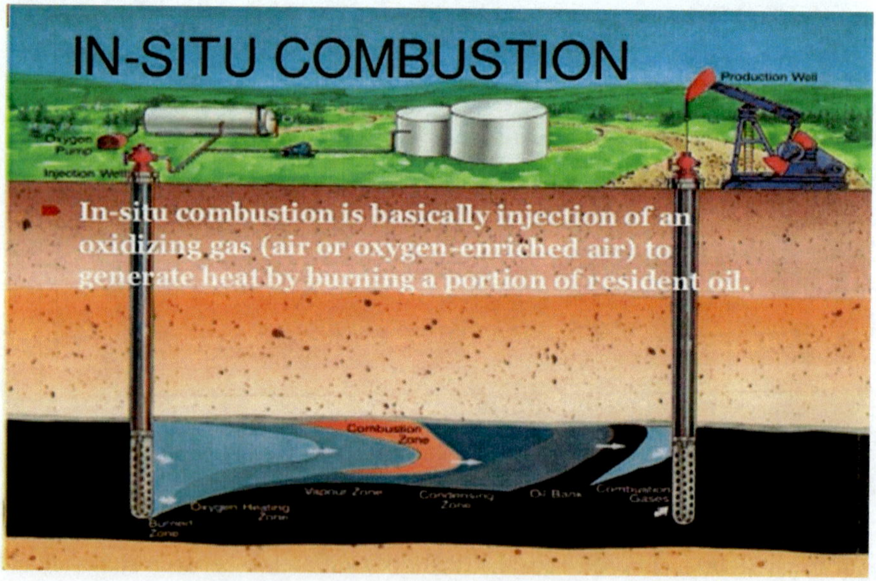

Fig. 7.4 Schematic diagram of insitu recovery of oil shale

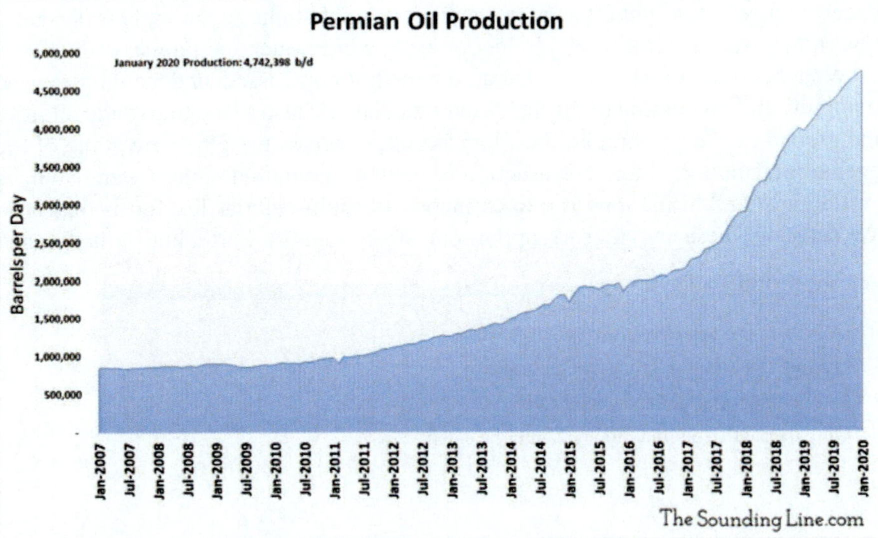

Fig. 7.5 US shale oil production in the Permian Basin. *Source* The Sounding Line, January 2020

the U.S. (Fig. 7.5) mainly for export (Katakay 2019). However, Reuters (July 2020) reports current reduction in shale oil production due to world-wide lower demand for oil.

7.2 The Economics of Oil Produced from Oil Shales

The cost of producing oil from **oil shales** in countries like Estonia as well the currently produced oil and gas from **shale oil** in Permian Basins in the US are uneconomical, however are in production only by being heavily subsidised. Shale oil/gas production cost between $40 and $60 to produce a barrel. Studies by the Stockholm Institute and Earth Track, Inc. (Mikulka 2017) found that '*at a price of $50 per barrel, light oil produced by hydraulic fracturing ("fracking") was heavily dependent on subsidies*'. The same economics apply to oil shales, as oil produced from **oil shale** is even higher in cost than from **shale oil** as it involves substantial energy input into its production and upgrading before it can go to an oil refinery. It is difficult to get an accurate price due to different techniques and different retorts used and very different labour cost. However, it has been reported by Craig (2011) based on a survey conducted by RAND Corporation that the cost of oil from oil shale at a model surface retorting complex with a retorting plant, upgrading plant, supporting utilities and spent shale reclamation Price may range between US$70 and US$95 a barrel. Estonia's oil shale industry is heavily subsidised by that country's government. It is however their only natural energy resource. The same goes for Jordan that has no other natural resources, except plenty of sunlight for solar power generation. Estonia accordingly is presently investing in renewable energy production from wind, water and biomass.

What becomes quite clear is that all the methods and associated techniques used in mining and processing of oil shales in particular, but also extracting crude oil from the ground and sea floor, and their transportation across the globe are some of the agents contributing to mass destruction of habitats, contamination of water sources and oceans, and in the long run to extinction of multi-cellular life forms on Earth. We carry out these activities to support our life systems on Earth, but for how long?

A Cree prophecy says:

> Only after the last tree has been cut down,
> Only after the last river has been poisoned
> Only after the last fish has been eaten
> Only then will you find money cannot be eaten.

References

Craig, J. H. S. (2011). Oil shales processing and the environment (16p). In *8th International Conference on Environmental Engineering*.

Dixon, D. A., & Pope, G. (1987). Oil shale of the Duaringa Basin, Central Queensland. *Fuel, 66*, 305–308.

Frank, D. (2000). *Public Submission-Stuart oil shale project* (28p).

Glikson, M., Chappell, B. W., & Webber, E. (1987). Trace element association in raw and spent shales from Rundle oil shale complex in Queensland. *Fuel, 66*, 400–403.

Glikson, M., Golding, S. D., Boreham, C. J., & Saxby, J. D. (2000). Mineralization in eastern Australian coals: A function of oil generation and primary migration. In M. Glikson & M. Mastalez (Eds.), *Organic matter and mineralization: Thermal alteration, hydrocarbon generation and role in metallogenesis* (pp. 329–358). Netherlands: Kluwer Academic Publishers.

Grunewald, E. (2006). *Oil shale and the environmental cost of production* (pp. 1–16). California: Stanford University.

Katakay, (2019). *Oil 2019*; IEA Report.

Koel, M. (1999). *Estonian oil shale* (18p). Compiled Report.

Lotman, S. (2016). *Don't let Estonian oil from shale do to Utah what it has done to Estonia* (4p). The Salt Lake Tribune, June 2016.

Mikulka, J. (2017). *Welfare kings? Study find half of new oil production unprofitable without government handouts* (5p). DESMOG.

Pope, G. J. (2000). The application of sequence stratigraphy in modelling oil yield in the stuart oil shale deposit. M.Sc. Thesis, QUT, published. (146 pp).

Raukas, A., & Punning, J.-M. (2009). Environmental problems in Estonian oil shale industry. *Royal Society of Chemistry—Energy and Environmental Science, 2*, 723–728.

Schmidt, S. J. (2003). New directions for shale oil: Path to a secure new oil supply well into this century. *Oil Shale, 20*(3), 333–346.

Appendix
Details of Samples Used in This Study

Sample no	Borehole, core	Lithology	Main minerals (XRD)
Toolebuc formation			
L6-1	Longreach No. 6 Core 1; 25.00 m	Dark grey siltstone with shells (*Inoceramus*)	
L6-2	Longreach No. 6 Core 1; 25.30 m		
L6-3	Longreach No. 6 Core 1, 25.60 m		
L6-4	Longreach No. 6 Core 2, 28.50 m	Black shale with shells (*Inoceramus*, *Oucelina*)	Illite, calcite, quartz, pyrite
L6-5	Longreach No. 6 Core 2; 29.00 m		
L6-6	Longreach No. 6 Core 1; 25.65 m		
L6-7	Longreach No. 6 Core 1; 25.80 m		
L6-8	Longreach No. 6 Core 1; 25.90 m	Black shale with shells	
L6-9	Longreach No. 6′ Core 1; 25.95 m		
L6-10	Longreach No. 6 Core 1; 27.50 m		
L6-11	Longreach No 6 Core 2; 31.00 m		
L6-12	Longreach No. 6 Core 1; 28.00 m	Calcareous shale	
J11-1	Jericho 11, Core 1 44.20 m	Grey-green mudstone	
J11-2	Jericho 11 Core 1; 45.80 m	Black clayey shale, very fine laminations	

(continued)

M. Glikson-Simpson, *Oil Shales*, https://doi.org/10.1007/978-3-030-60675-6

(continued)

Sample no	Borehole, core	Lithology	Main minerals (XRD)
J11-3	Jericho 11 Core 1; 44.50 m		
J11-4	Jericho 11 Core 2; 45.50 m	Black clayey shale, very fine laminations	
J11-5	Jericho 11 Core 3; about 50 m		
J11-6	Jericho 11 Core 3; 50.50 m	Dark grey laminated shale	
J11-7	Jericho 11 Core 4; 51.50 m		
J11-8	Jericho 11 Core 4; 54.00 m	Dark grey claystone	
J11-9	Jericho 11 Core 4; 53.90 m	Finely laminated black shale Some calcareous laminae	
J11-10	Jericho 11 Core 5; 54.30 m	Dark grey, finely laminated shale	
J11-11	Jericho 11 Core 6; 57.50 m	Finely laminated black shale	
J11-12	Jericho 11 Core 7; 61.00 m	Finely laminated dark grey shale	
J11-13	Jericho 11 Core 7; 62.00 m		
J11-14	Jericho 11 Core 8; 63.50 m		
J11-15	Jericho 11 Core 8; 64.00 m	Finely laminated Black shale with calcareous Lenses, *Inoceramus* remains	
J11-16	Jericho 11 Core 8; 64.50 m		
J11-17	Jericho 11 Core 8; 65.50 m	Finely laminated black shale	
J11-18	Jericho 11 Core 8; bottom		
J11-19	Jericho 11 Core 9; 67.00 m	Light grey siltstone	
J11-20	Jericho 11 Core 9, bottom	Dark grey shale	
J11-21	Jericho 11 Core 9, 68 m		
J11-22	Jericho 11 Core 10; 69.50 m	Light grey claystone	

(continued)

(continued)

Sample no	Borehole, core	Lithology	Main minerals (XRD)
J11-23	Jericho 11 Core 10; 70.00 m	Light grey claystone with carbonised remains	
J11-24	Jericho 11 Core 11; 71.00 m	Finely laminated calcareous black shale	
J11-25; J11-26	Jericho 11 Core 11; 71.00 m	Light grey sandy siltstone limonitic patches, calcareous	
3-3	Tambo 37 Core 3; 68.00 m	Dark grey mudstone	Quartz, illite, montmorillonite
3-4 to 9-1	Tambo 37 About 69 m		Kaolinite, albite, quartz
2-1 10-1 10.2 6-1, 6-2	Tambo 38 Core 1	Black calcareous shales, finely laminated	
2-2	Tambo 38 Core 2	Dark grey claystone	Montmorillonite Kaolinite, albite Quartz
1-1, 10-3, 4 11-1	Tambo 38 Core 3; 57.20 m	Black shale with shells	Illite, Montmorilo. Quartz, calcite Chlorite, pyrite
1-3; 1-2 11-2	Tambo 38 Core 4		
6-3	Tambo 38 Core 5		
11-3;1-3			
2-3	Core 6	Black shale	Kaolinite, quartz Montmorilonite
1-4, 12-3	Core 7	Dark grey siltstone	Montmoril. Calcilte
2-4 9-3 9-4	Tambo 38 Core 8	Light grey siltstone carbonised remains light and dark grey mudstone	Montmorill. Phosphor-pherrite
B3-1	Boulia 3 105′7″–107′	Black shale, calcareous lenses	
B3-2	Boulia 3 92′–99′		
B3-3	Boulia 3 100′11″–101′		
B3-4	Boulia 3 108′4″–109′3″		
B3-5	Boulia 3 104′4″–105′3″	*Inoceramus* abundant	
B3-6	Boulia 3 109′3″–110′2″		

(continued)

(continued)

Sample no	Borehole, core	Lithology	Main minerals (XRD)
B3-7	Boulia 3 112′8″–113′6″		
B3-8	Boulia 3 120′–122′	Grey-greenish mudstone	
T/13/14	Tambo 42 Core 1	Dark grey claystone with carbonised organic remains	
T/15–T/17	Tambo 42 Core 2		
T/18–T/19	Tambo 42 Core 3		
T/20	Tambo 42 Core 4	Grey siltstone	
AU5-51, 52, 53	Augathella 5 Core 1	Dark grey-black mudstone with coaly lenses	
AU5-54	Augathella 5 Core 2	Grey clayey mudstone	
AU5-55	Core 3	Grey clayey sandstone	
AU6-40	Augathella 6 Core 1; 16.18–16.50 m	Black-grey siltstone	
AU7-22, 23	Augathella 7 Core 19–20	Black shale with calcareous matter	
AU7-24	Augathella 7 Core 20		
AU7-25	Augathella 7 Core 21		
AU7-26, 27	Augathella 7 Core 22		
GS-1	Birigoolpa WRC-30961 430′ (cuttings)	Light grey silty claystone	
GS-2	Birigoolpa WRC-30961 415′ (cuttings)	Dark grey clayey siltstone	
GS-3	Birigoolpa WRC-30961 400′ (cuttings)		
GS-8	Birigoolpa WRC-30961 389′ (cuttings)	Sandstone with coaly particles	

(continued)

(continued)

Sample no	Borehole, core	Lithology	Main minerals (XRD)
GS-4, GS-7	Urella Downs WRC-30944 360′ (cuttings)	Grey siltstone with coaly particles	
GS-6	Urella Downs WRC-30944 98–99′ (cuttings)	Claystone	
Lower palaeozoic			
C-1	Mt Isa-1 160.35 m	Hard, compacted black shale with thin calcareous bands	Illite, chlorite Calcite, dolomite, albite
C-2	Mt Isa-1 83.55 m	Compacted black shale with thin calcareous bands	Illite, chlorite, calcite Quartz
C-3	Mt Isa-1 68.7 m		
C-4	Mt Isa-1 82.88 m	Black chert	Quartz
C-5	Mt Isa-1 92.40 m	Finely laminated, calcareous black shale	
C-6	Mt Isa-1 95.52 m		
C-7	Mt Isa-1 107.60 m		
C-8	Mt Isa-1 114.50 m		
C-9	Mt Isa-1 117.43 m		
C- 17	Mt Isa-1 50.19 m		
C-12	Mt Isa-1 117.46 m		
C-13	Mt Isa-1 123.81 m		
C-21	Mt Isa-1 135.26 m		
C-22	Mt Isa-1 141.76 m		
C-23	Mt Isa-1 185.62 m	Calcareous black shales	
C-24	Mt Isa-1 215.73 m	Limestone with organic bands	
C-18	Camooweal-2 108.78 m	Limestone with organic lenses	

(continued)

(continued)

Sample no	Borehole, core	Lithology	Main minerals (XRD)
C-20	Camooweal-2 77.79 m	Calcareous shale	
RO4-10	Rodinga-4 Core 31 106.00 m	Black highly compacted shale with calcareous lenses	
RO-11	Rodinga-4 Core 36 119.79 m		
RO-12	Rodinga-4 Core 28 97.06 m	Black shales with calcareous laminae	
PS	Thorntonia limestone	Black organic-rich band overlying glauconitic horizon	
By-1	Byalkaoora-1 702–735′	Black shale with halite crystals and dolomite	
The Ethabuca borehole samples			
Depth ft.	*Type of sample*		
4010′	Cuttings	Devoid of OM	
5080′		OM detrital of high Ro; solidified immiscible heavy oil	
5470′		Poor in OM, detrital pyrobitumen	
5690′		Abundant pyrobitumen between carbonate, pyrite	
5820′		OM lamellar between carbonate, pyrite	
6420′		Only detrital bitumen	
The Brothers Borehole samples			
1245	Core	Carbonate veins, lenses and bands impregnated with pyrobitumen	
1420		No OM	
1588		Pods and lenses of pyrobitumen in calcareous rock representing fossil oil reservoir	
1952		OM lamellar after oil generation. Dispersed Galena	

(continued)

(continued)

Sample no	Borehole, core	Lithology	Main minerals (XRD)
2060	Cuttings	Several generations of OM	
2235		Devoid of OM, sulphides	
2403	Core		
2414			
2896		Bands and lamellae of OM	
3432		No OM	
3912		OM residue after oil generation, bitmen and pyrobitumen	
4030	Cuttings	High in OM; 3 types identified within carbonates	
4060		High in OM. Pyrite	
4154	Core	Pyrobitumen in carbonate. Possibly residual oil in reservoir	
Tertiary oil shales			
R-1	Rundle oil shale sequence	Light brown shale	Illite, montmorill., pyrite quartz
Con-1	Condor oil shale	Algal, brown oil shale	
Con-2		Black lignite	
GR-1	Green River, Lake Gosiute Mine	Pure organic matter concentrate	
GR-2		Brown oil shale with dolomite and shortite, coarse grained	
GR-3		Grey-greenish fine-grained kaolinite oil shales	Kaolinite
Mount coolon			
183/1	Mt Coolon 183 54.9–55.3 m	Lignite	
183/2		Kaolin impregnated with organic matter	Kaolinite, quartz
183/3	56.1–56.3 m	Brown oil shale	
183/20	51.77–52.37 m	Lignite and pyrite	Pyrite
183/21	53.10–53.53 m	Brown, very loosely packed oil shale	
183/22	53.53–53.86 m	Brown oil shale	
183/23	53.83–55.30 m	Lignite and pyrite	

(continued)

(continued)

Sample no	Borehole, core	Lithology	Main minerals (XRD)
183/24	55.3–55.6 m	Brown oil shale	
283/4	Mt Coolon 283 39.6–40.1 m		
283/5	Mt Coolon 283 43.6–44.3 m	Brown oil shale with massive pyrite	
283/6	Mt Coolon 283 52.7 m	Brown oil shale with coaly patches	
283/7	Mt Coolon 283 52.7–53.3 m	Brown oil shale merging into lignite	
283/8	Mt Coolon 283 52.3–55.7 m	Brown oil shale with coaly parts	
283/9	Mt Coolon 283 56.1–56.6 m	Brown waxy oil shale	
283/10	Mt Coolon 283 57.7–58.1 m		
283/11	Mt Coolon 283 59.2–59.6		
283/12	Mt Coolon 283 58.8 m	Dark brown woody lignite	
283/13	Mt Coolon 283 60.2–60.7 m	Brown waxy oil shale with resin nodules	
283/14	Mt Coolon 283 60.7–61.1 m	Dark brown oil shale with woody remains	
283/15	Mt Coolon 283 61.6–62.0 m	Brown oil shale	
283/16	Mt Coolon 283 62.1–62.6 m	Brown waxy shale, wood remains	
283/17	Mt Coolon 283 63.5 m	Brown waxy oil shale	
283/18	Mt Coolon 283 64.0 m	Brown oil shale with coaly patches	
283/19	Mt Coolon 283 64.5 m	Black lignite	
COO-1	Mt Coolon RDH-70B 46 m	Brown oil shale with coaly patches	
COO-2	Mt Coolon 70B 50 m		
COO-3	Mt Coolon 70B 53 m		
COO-4	Mt Coolon 70B 55 m	Brown oil shale with pyrite nodules	

(continued)

(continued)

Sample no	Borehole, core	Lithology	Main minerals (XRD)
COO-7	Mt Coolon 70B 57 m	Brown oil shale with coaly remains	
COO9A-5	Mt Coolon 9A } 48–50 m	Oil shake and lignite (mixed) weathered samples	
COO9A-3	Mt Coolon 9A } 48–50 m		
COOA-2	Mt Coolon 9A } 48–50 m		
Permiam oil shales			
TA-1	Torbanite, Alpha	Black silty torbanite, brittle	
TA-2		Black hard, conchoidal fracture	
TC-1	Torbanite, Carnarvon Creek, Queensland		
T-GD	Torbanite, Glen Davis NSW		
T-M	Torbanite, Marangaroo		
T-S	Torbanite, Cessnock, NSW	Black, hard, conchoidal fracture	
T-1	Tasmanite, Mersey River, Tasmania	Dark grey, hard shale, white patches	
Other oil shales			
K-1	Kukersite, Estonia	Reddish matrix with numerous white patches. Very hard	
COO-1	Coorongite, Coorong Lagoon, South Australia	Light grey-greenish silty	
COO-2	Coorongite, Porrarecup lagoon, Western Australia	Black-greenish rubbery accumulation of *Botryococcus* alga	

Index

M. Glikson-Simpson, *Oil Shales*, https://doi.org/10.1007/978-3-030-60675-6